# The Technological Competitiveness of Japanese Multinationals

**Thames Essays on Contemporary International Economic Issues**

Published under the aegis of the Trade Policy Research Centre, London

Books in this series focus on key topics affecting the international economy, including trade, the activities of multinational enterprises, and international macroeconomics. They are written by leading academics and practitioners in international economics and addressed to the business and policy-making communities as well as the academic market. Each essay covers an issue of topical concern.

Titles in the Series:

*The Technological Competitiveness of Japanese Multinationals,* by Robert Pearce and Marina Papanastassiou

# The Technological Competitiveness of Japanese Multinationals

## The European Dimension

Robert Pearce
and Marina Papanastassiou

*Ann Arbor*
THE UNIVERSITY OF MICHIGAN PRESS

Published in the United States of America by
The University of Michigan Press
Manufactured in the United States of America
♾ Printed on acid-free paper

1999 1998 1997 1996 4 3 2 1



*A CIP catalog record for this book is available from the British Library.*

Library of Congress Cataloging-in-Publication Data

Pearce, R. D. (Robert D.)
The technological competitiveness of Japanese multinationals : the European dimension / Robert Pearce and Marina Papanastassiou.
p. cm. — (Thames essays)
Includes bibliographical references and index.
ISBN 0-472-10728-3
1. International business enterprises—Japan. 2. Corporations, Japanese—Europe. 3. High technology industries—Europe. 4. Electronic industries—Europe. 5. Research, Industrial—Europe. 6. Competition, International. I. Papanastassiou, Marina. II. Title. III. Series.
HD2907.P4 1996
338.8′8895204—dc20 96-4456
CIP

Στη γιαγιά μου Χρυσάνθη Λουράντου

## Acknowledgments

We are very grateful to John Cantwell for providing the patent data used in this book. We would also like to thank the secretaries who typed first drafts of various chapters and, especially, Yvonne Penford for her usual expert production of the final manuscript.

## Contents

Chapter 1. The Rise of Japanese Multinationals and Technology 1

Chapter 2. Internationalized Research and Development 27

Chapter 3. Internationalization of Technology in Japanese Multinationals 49

Chapter 4. Japanese MNEs' R&D Operations in Europe 59

Chapter 5. Japanese R&D Laboratories in the United Kingdom 75

Chapter 6. Japanese Electronic MNEs' Subsidiaries in Europe 85

Chapter 7. Conclusions 95

References 101

Index 105

CHAPTER 1

# The Rise of Japanese Multinationals and Technology

The past decade has seen an enormous increase in the number of production operations in Europe owned by Japanese multinational enterprises (MNEs). A familiar interpretation of this rise is to see it as being essentially defensive, with Japanese production in Europe replacing imports as a response to actual or anticipated protectionist measures. The argument of this book indicates that the European manufacturing activity of Japanese MNEs can be seen in a much more positive light, frequently involving the development of distinctive European products (rather than merely supplying locally the established goods that would otherwise have been imported) as part of a wider global strategy.[1] The increasingly creative orientation of Japanese companies' manufacturing operations in Europe is reflected in a massive rise in the number and scope of research and development (R&D) facilities in the region that are controlled by these MNEs. These R&D units may contribute to both the creation of new scientific knowledge and its application as new European products in ways that often amount to playing a role in a global-innovation strategy.

The emergence of a globalized approach to innovation in Japanese MNEs can be perceived as a logical response to the interaction of two key trends in the nature and behavior of these companies.[2] The first of these trends has been the increasing extent and geographical dispersion of their overseas production. Whatever the original motivation for the extension of production into particular countries may have been, a normal outcome is greater knowledge of the distinctive characteristics of their markets and greater motivation within the local subsidiaries to seek permission to try to satisfy these specific original needs. A sensitive MNE should therefore be becoming ever more aware of an increasing range of decentralized market signals. Alongside this implication

1. Even when European production does substitute for imports, the implementation of this "localized" supply is usually through an efficient, intra-European-trade-oriented strategy integrating networks of specialized subsidiaries. Thus the vast majority of Japanese manufacturing subsidiaries in Europe seem to supply the European rather than national markets. See Papanastassiou and Pearce 1993, 1994c.

2. See Papanastassiou and Pearce 1994a, 155–56.

of rising overseas production, the second key trend in the Japanese companies is an ever growing commitment to technology as a central element in their competitiveness. Having moved through phases of cost-efficient production of existing standardized products, and then high-quality product differentiation within established ranges, many Japanese companies have now joined those seeking to compete at the leading edge of the creation of vital new product concepts and generations. These complementary competitive objectives, of introducing significant new products and responding effectively to the distinctive characteristics of important regional/national markets, imply the implementation of a globalized approach to innovation itself.

The background to a global-innovation strategy that we have just outlined allows us to discern two sets of factors that determine the roles that may be played by overseas R&D laboratories. First, what may be described as demand-side factors will clearly be vital in determining the nature of the operations of many such labs. In many industries, rising income levels do not lead to increasingly homogeneous global markets; on the contrary, it is often the case that greater prosperity provides consumers with greater scope to manifest original elements in their tastes. Even where the innovation of major new products is a key element in competition, therefore, their uniqueness is, of itself, likely to provide only a limited and temporary competitive edge, and the ability to enhance this edge, right from the original implementation, by responsiveness to the distinctive characteristics of various important markets may be vital. Thus overseas R&D laboratories may develop original variants of the new product that fully embody the unique elements of consumer tastes, and often local production conditions, of their regional (e.g., European) or national markets.

The second set of factors that may motivate the establishment of overseas R&D laboratories in Japanese MNEs, and help determine their roles, may be described as supply-side influences and relate to the quality and areas of specialization of the scientific capabilities in particular host countries. In very general terms, Japanese firms may be especially willing to respond to the availability of such dispersed scientific inputs due to a perceived weakness of the Japanese research capacity in some significant areas of technology. A more detailed influence that these Japanese companies may share with other MNEs relates to the widening range of technologies that are now embodied in most major innovations. With their home-country science base frequently unable to supply high-quality inputs in all of the relevant areas of expertise, MNEs are increasingly using their growing knowledge of worldwide scientific specializations to access effectively the relevant selection of capabilities. Thus some Japanese companies' laboratories outside of Japan may be tapping into the particular specialized research capability of their host country's science base in order to assist in the group's aim of assembling the best available supplies in all the relevant areas of technology needed to create the new background scientific knowledge from which the commercial innovations can be derived.

The two remaining sections of this chapter investigate the expansion of Japanese companies' overseas production and growing commitment to leading-edge technology, which we have argued provide the background to the emergence of the global approach to innovation that embodies decentralized R&D. The second chapter then discusses in more detail the concept of a global-innovation strategy and the roles of overseas R&D laboratories. The third chapter documents the rise of Japanese MNEs' decentralized R&D operations worldwide, and the fourth then focuses in much greater detail on the extent, motivation, and roles of their labs in Europe. Chapter 5 reports the results of an original survey of Japanese MNEs' laboratories in the United Kingdom, while chapter 6 reviews evidence from another survey covering the operations of Japanese electronics industry subsidiaries in Europe. Chapter 7 provides a summary and conclusions.

## The Rise of Japanese Firms' Operations in Europe

In the first part of this section we review evidence on stocks and flows of foreign direct investment (FDI) that illustrate the substantial recent rise in the relative contribution of Japanese enterprises to this important facet of the global economy. We also discuss some theorizing relating to the changing motives for, and roles of, the overseas activities of such Japanese MNEs. In the second part we scrutinize data on various aspects of Japanese MNEs' operations in Europe and look at some evidence on the factors that have motivated their recent growth.

### The Rise of Japanese MNEs

Data provided by JETRO, contained in table 1.1, allows us to discern several crucial trends in the geographical patterns of FDI in recent years. The first trend we can note from these FDI flow figures is indeed the rise in importance of the outflows of such investments from Japan. Thus, of the total outflows of FDI from Japan, the United States, and European Community (EC) countries, Japanese companies' investments accounted for 7.2 percent in 1979–81 but for 27.3 percent in 1988–90. Data on stocks of FDI compiled by the UNCTAD Program on Transnational Corporations confirms the rise of Japan in outward FDI, indicating that its companies increased their share of the total stocks of outward FDI from 7.1 percent in 1980 to 19.3 percent in 1990 (table 1.2). This meant that by 1990 Japan was clearly the second most important outward investor, after the United States, in terms of asset stocks, having also

**TABLE 1.1. Outflows of Foreign Direct Investment from Japan, United States, and European Community, 1979–90**

| | 1979–81 | | 1982–84 | | 1985–87 | | 1988–90 | |
|---|---|---|---|---|---|---|---|---|
| | $ mln | % | $ mln | % | $ mln | % | $ mln | % |
| Japan | | | | | | | | |
| To U.S. | 3,620 | 35.6 | 6,031 | 42.7 | 20,171 | 49.9 | 65,791 | 52.1 |
| To EC | 1,101 | 10.8 | 1,982 | 14.0 | 7,876 | 19.5 | 26,566 | 21.0 |
| To rest of world | 5,456 | 53.6 | 6,104 | 43.2 | 12,404 | 30.7 | 34,007 | 26.9 |
| Total | 10,177 | 100.0 | 14,117 | 100.0 | 40,451 | 100.0 | 126,364 | 100.0 |
| U.S. | | | | | | | | |
| To Japan | 1,306 | 2.5 | 1,106 | 5.7 | 2,623 | 4.2 | 3,180 | 3.8 |
| To EC | 23,313 | 44.7 | 11,653 | 60.5 | 23,844 | 37.9 | 35,848 | 42.3 |
| To rest of world | 27,555 | 52.8 | 6,490 | 33.7 | 36,419 | 57.9 | 45,676 | 53.9 |
| Total | 52,174 | 100.0 | 19,249 | 100.0 | 62,886 | 100.0 | 84,704 | 100.0 |
| EC | | | | | | | | |
| To Japan | 152 | 0.2 | 155 | 0.3 | 226 | 0.2 | 1,533 | 0.6 |
| To U.S. | 25,681 | 32.3 | 30,515 | 51.6 | 70,687 | 54.7 | 84,433 | 33.5 |
| Intra-EC[a] | 8,193 | 10.3 | 7,830 | 13.2 | 23,961 | 18.5 | 66,542 | 26.4 |
| To rest of world | 45,527 | 57.2 | 20,624 | 34.9 | 34,434 | 26.6 | 99,261 | 39.4 |
| Total | 79,553 | 100.0 | 59,124 | 100.0 | 129,308 | 100.0 | 251,769 | 100.0 |
| Share by source | | | | | | | | |
| Japan | 10,177 | 7.2 | 14,117 | 15.3 | 40,451 | 17.4 | 126,364 | 27.3 |
| U.S. | 52,174 | 36.8 | 19,249 | 20.8 | 62,886 | 27.0 | 84,704 | 18.3 |
| EC | 79,553 | 56.1 | 59,124 | 63.9 | 129,308 | 55.6 | 251,769 | 54.4 |
| Total | 141,904 | 100.0 | 92,490 | 100.0 | 232,645 | 100.0 | 462,837 | 100.0 |

*Source:* Calculated from JETRO 1992a, table 3, 18.

[a]Intraregional foreign direct investment in EC in 1984–90 represents figures in the United Kingdom, Germany, France, Italy and the Netherlands.

ranked below the United Kingdom, Germany, and the Netherlands in 1980. Additional JETRO data on the geographical distribution of Japanese FDI flows (see table 1.3) indicates that Europe and the United States have both increased their shares, mainly at the expense of investments in Asia. In particular, the recent strength of investment in Europe is confirmed, showing that it claimed 21.8 percent of all Japanese flows in 1987–92 compared with 14.4 percent in 1981–86 and 12.3 percent in 1951–80. By contrast with this pronounced emergence of Japan as an outward investor, table 1.1 illustrates the persistence of its status as a minor recipient of inward investment, with only 1.4 percent of the total FDI flows of the United States and EC countries in 1988–90 going to it. The UNCTAD stock data confirms this situation, with Japan accounting for only 1.3 percent of total recorded inward FDI stocks in 1990 compared with 0.7 percent in 1980.[3]

3. For discussion of this see UNCTAD 1993, 42–43.

**TABLE 1.2. Stock of Foreign Direct Investment, by Investing Country, 1980, 1985, 1990**

| | Value ($ bln) | | | Share (%) | | |
|---|---|---|---|---|---|---|
| | 1980 | 1985 | 1990 | 1980 | 1985 | 1990 |
| European Commmunity (total) | 194.5 | 254.3 | 666.4 | 37.7 | 36.8 | 41.3 |
| Belgium and Luxembourg | 6.0 | 4.7 | 22.7 | 1.2 | 0.7 | 1.4 |
| France | 12.2 | 20.3 | 74.8 | 2.4 | 2.9 | 4.6 |
| Germany | 43.1 | 59.9 | 155.1 | 8.4 | 8.7 | 9.6 |
| Italy | 7.0 | 16.3 | 56.1 | 1.4 | 2.4 | 3.5 |
| Netherlands | 42.1 | 47.6 | 87.3 | 8.2 | 6.9 | 5.4 |
| United Kingdom | 80.7 | 101.2 | 244.8 | 15.7 | 14.7 | 15.2 |
| Other Western Europe (total) | 31.4 | 43.1 | 89.8 | 6.1 | 6.2 | 5.6 |
| Sweden | 8.0 | 13.9 | 24.9 | 1.6 | 2.0 | 1.5 |
| Switzerland | 21.5 | 25.3 | 52.6 | 4.2 | 3.7 | 3.3 |
| North America (total) | 242.8 | 289.8 | 497.9 | 47.1 | 42.0 | 30.9 |
| United States | 220.2 | 251.0 | 423.2 | 42.7 | 36.4 | 26.2 |
| Canada | 22.6 | 38.7 | 74.7 | 4.4 | 5.6 | 4.6 |
| Other Developed Countries (total) | 44.7 | 97.6 | 348.1 | 8.7 | 14.1 | 21.6 |
| Australia | 2.3 | 6.7 | 27.4 | 0.4 | 1.0 | 1.7 |
| Japan | 36.5 | 83.6 | 310.8 | 7.1 | 12.1 | 19.3 |
| Developed Countries (total) | 513.3 | 684.9 | 1602.2 | 99.6 | 99.2 | 99.3 |
| Developing Countries (total) | 2.2 | 5.4 | 10.8 | 0.4 | 0.8 | 0.7 |
| Total Outward Investment | 515.5 | 690.2 | 1613.0 | 100.0 | 100.0 | 100.0 |

*Source:* UNCTAD 1993, annex table 2, 248–50.

A second key trend in FDI in recent years, complementing the rise of Japan, is the relative decline of the United States as a source country. Thus table 1.1 shows that in 1988–90 U.S. firms accounted for only 18.3 percent of the total outward FDI from the United States, Japan, and the European Community (which ranked it a poor second to Japan), compared with 36.8 percent in 1979–81 (when it was still clearly established as the leading outward investor). The outward FDI stock figures of table 1.2 also reflect this decline, with the United States accounting for only 26.2 percent of all such stocks in 1990, compared with 42.7 percent in 1980. The United States was, however, also responsible for a third significant trend in recent FDI, in the form of its rising importance as a host country. Thus the UNCTAD data (UNCTAD, 1993, annex table 2) records that the United States accounted for 27.5 percent of all inward FDI stocks in 1990 compared with 18.5 percent in 1980. We have already drawn attention to the growing share of the United States in Japanese outward FDI flows, as shown in table 1.3, and it can be observed that table 1.1 confirms this growth. In fact the data of table 1.1 shows that the $65,791 million of Japanese FDI in the United States in 1988–90 accounted for 43.8 percent of the total of $150,224 million undertaken there by Japanese

**TABLE 1.3. Geographical Distribution of Japanese Direct Investment Abroad—Value**

| | Value, U.S. $ million (cumulative) | | | Share of Value (percent) | | |
|---|---|---|---|---|---|---|
| | 1951–80 | 1981–86 | 1987–92 | 1951–80 | 1981–86 | 1987–92 |
| United States | 8,878 | 26,577 | 126,918 | 24.3 | 38.3 | 45.2 |
| Canada | 920 | 1,031 | 5,256 | 2.5 | 1.5 | 1.9 |
| North America (total) | 9,798 | 27,608 | 132,174 | 26.8 | 39.7 | 47.1 |
| United Kingdom | 2,009 | 2,116 | 25,009 | 5.5 | 3.0 | 8.9 |
| Netherlands | 298 | 2,039 | 13,885 | 0.8 | 2.9 | 4.9 |
| Germany | 497 | 1,055 | 5,022 | 1.4 | 1.5 | 1.8 |
| France | 384 | 586 | 4,459 | 1.1 | 0.8 | 1.6 |
| Other Europe | 1,284 | 4,203 | 12,851 | 4.6 | 6.0 | 4.6 |
| Europe (total) | 4,472 | 9,999 | 61,226 | 12.3 | 14.4 | 21.8 |
| Australia | 2,164 | 2,338 | 16,261 | 5.9 | 3.4 | 5.8 |
| Oceania (total) | 2,525 | 2,709 | 18,548 | 6.9 | 3.9 | 6.6 |
| Indonesia | 4,424 | 4,249 | 5,736 | 12.1 | 6.1 | 2.0 |
| Hong Kong | 1,095 | 2,338 | 8,077 | 3.0 | 3.4 | 2.9 |
| Singapore | 936 | 1,635 | 5,266 | 2.6 | 2.4 | 1.9 |
| Thailand | 396 | 488 | 5,003 | 1.1 | 0.7 | 1.8 |
| Malaysia | 650 | 633 | 3,532 | 1.8 | 0.9 | 1.3 |
| Rep. of Korea | 1,136 | 982 | 2,505 | 3.1 | 1.4 | 0.9 |
| China | — | 514 | 3,958 | — | 0.7 | 1.4 |
| Taiwan | 370 | 682 | 2,375 | 1.0 | 1.0 | 0.8 |
| Asia (total) | 9,830 | 11,960 | 38,090 | 26.9 | 17.2 | 13.6 |
| Middle East (total) | 2,259 | 758 | 1,214 | 6.2 | 1.1 | 0.4 |
| Africa (total) | 1,445 | 2,234 | 3,133 | 4.0 | 3.2 | 1.1 |
| Panama | 687 | 8,154 | 9,898 | 1.9 | 11.7 | 3.5 |
| Brazil | 2,908 | 1,949 | 2,338 | 8.0 | 2.8 | 0.8 |
| Latin America (total) | 6,168 | 14,205 | 26,174 | 16.9 | 20.4 | 9.3 |
| Total | 36,497 | 69,473 | 280,560 | 100.0 | 100.0 | 100.0 |

*Source:* JETRO (mimeo documentation).
*Note:* Calculations are based on approval basis up to November 1980 and on reported basis since then.

and EC ($84,433 million) companies during that period. The figure for 1979–81, calculated in the same fashion, was only 12.4 percent.[4] The United States also received an increase in the outward FDI of EC firms once intra-EC investments are excluded. Thus in 1978–81 EC investments in the United States were $25,681 million, which accounted for 36.0 percent of the total of $71,360 million undertaken outside of the EC by EC companies. The comparable figure for 1988–90 was 45.6 percent (i.e., $84,433 million from total extra-EC investment of $185,227 million).

4. The figures for the intervening periods were 16.5 percent in 1982–84 and 22.2 percent in 1985–87.

The major trend involving the EC countries that is depicted in table 1.1 is the rise of intra-EC FDI, this accounting for 10.3 percent of the total outward FDI of EC companies in 1979–81 and for 26.4 percent in 1988–90. Looked at from the alternative perspective, intra-EC FDI of $8,193 million in 1979–81 accounted for 25.1 percent of the total inward FDI of $32,607 million received by the EC countries from the three sources covered in table 1.1 (i.e., $8,139 million intra-EC, $1,101 million from Japan, and $23,313 million from U.S.). The comparable proportions can be calculated as 36.5 percent in 1982–84, 43.0 percent in 1985–87, and 51.6 percent in 1988–90. If intra-EC FDI is excluded from the calculation, the data of table 1.1 reveals that the outward FDI of the EC countries ($71,360 million) in 1979–81 accounted for 53.4 percent of the total outward FDI of $133,711 million carried out by the United States ($52,174 million), Japan ($10,177 million), and the European Community in those years. This proportion can be calculated to have risen to 70.0 percent in 1982–84 (suggesting that European firms had a relatively less negative reaction, in terms of FDI, to the early 1980s recession than the United States in particular), but then to have fallen back to 50.5 percent in 1985–87 and to 46.7 percent in 1988–90. In terms of FDI stocks, the UNCTAD data produced in table 1.2 shows that the share of Europe as a whole in outward investment (including the intra-European component) was slightly higher in 1990 (at 46.9 percent) than it had been in 1980 (43.8 percent). Europe's strengthening position as a recipient of Japanese FDI, that we have already observed from table 1.3, is further reflected quite notably in results that can be derived from table 1.1. Thus, the Japanese FDI in European Community of $1,101 million in 1979–81 only amounted to 4.5 percent of the total of $24,414 million carried out in the community by U.S. ($23,313 million) and Japanese companies in that period. This proportion, however, then rose steadily to 14.5 percent in 1982–84, 24.8 percent in 1985–87, and 42.6 percent in 1988–90. Even if the calculation is reworked to include the fast-growing intra-EC FDI among the sources covered, Japanese companies still recorded notable rises (from 3.4 percent, to 9.2 percent, 14.1 percent, and 20.6 percent in the four successive periods) in the total inward FDI received by EC countries from the three sources (United States, Japan, intra-EC).

The data we have just surveyed clearly indicates that as Japanese companies have increased their position in terms of the overall level of FDI, they have also substantially reoriented their investments to focus increasingly on operations in Europe and North America. This reorientation suggests that the forces stimulating the overseas production of goods and services in Japanese enterprises are not only increasing in intensity but have also been subject to changes in their nature. Thus Ozawa (1991a, 137–43; 1991b, 51–54; 1992, 17–18) has discerned clear, if obviously overlapping, phases in Japanese manufacturing firms' overseas investments. These phases in the emergence of Japan's overseas investment activities are perceived as being closely related to the path of industrial restructuring in the home economy.

The initial reconstitution of effective industrial activity in Japan, through the 1950s and into the 1960s, involved the expansion of a number of labor-intensive manufacturing industries benefiting from low wages. However, in a manner that would be predicted by the Heckshcr-Ohlin theory, this development led to an increasingly tightened labor market in Japan by the mid-1960s, with a sharp rise in wages compromising the competitiveness of these usually strongly export-based industries. To avoid these wage-inflationary consequences of labor shortages, "Japanese manufacturers actively shifted the production of standardized, low-technology, labor-intensive goods (such as textiles, toys, simple electrical appliances and sundries) to neighboring countries such as Taiwan, South Korea, Hong Kong, Thailand and other Asian economies where labor was abundant," in what Ozawa (1991a) terms "the 'elementary' stage of offshore production." This wave of overseas investments probably peaked in the early 1970s, though the factors that motivated it probably continued to have some influence within the wider operations of Japanese MNEs.

In the face of the potentially "self-obsolescing" nature of a purely low-cost-labor-based strategy, Japanese industry (from the late 1950s to the early 1970s) sought another conventional source of competitiveness in the scale-economies-based modernization of heavy and chemical industries (e.g., steel, aluminum, shipbuilding, petrochemicals, synthetic fibers). However, the effective long-term development of these industries proved vulnerable in Japan, due to the fact that the country possessed few of the natural resources that they required and because their expansion "soon led to the malignancies of pollution, congestion and ecological destruction as [they] tended to generate undesirable environmental externalities" (Ozawa 1991a, 141). In addition the pursuit of export markets to realize the scale economies central to the strategy began to cause trade frictions. Once again the decreasing viability of an industrial stage in Japan provoked a second phase of overseas investments. In what Ozawa terms "the 'Ricardo-Hicksian trap' stage of transnationalism," the objectives were to escape from the trap[5] by transferring overseas some of the resource-intensive and/or pollution-prone operations and to secure stable supplies of industrial raw materials.

In terms of the nature of industrialization in Japan, the escape from the

5. In this case Ozawa (1991a, 141) characterizes "the trap of Ricardo-Hicksian industrialism" as being "the sheer impracticality of continuing the expansion of such resource-intensive, pollution-prone industries in a geographically confined, space-constrained small country." For elaboration of the Ricardo-Hicksian trap, see Ozawa 1979, 65–70.

Ricardo-Hicksian trap involved a concerted move[6] toward "less resource-consuming, more pollution-free and knowledge-intensive industries." Led by electronics, automobiles, and machine tools, the third wave of Japan's industrialization emphasized R&D-intensive, assembly-based, higher-value-added goods (Ozawa 1991a, 142). The success of this wave soon led to a major expansion of exports in these industries, which in turn provoked the rise of protectionist measures in North America and Europe (often in the "new" forms of quotas and voluntary export restraints) targeted against imports from Japan. The response was for the successful firms to set up production facilities (initially "screwdriver" assembly operations, but later often more locally integrated) in these markets. Thus, in a clear sense, this third phase of FDI is again a reaction to a particular problem faced by the industries in Japan (albeit this time an exogenously generated one, rather than one endogenous to Japanese production). However, this time the overseas operations complement strong continued activity in the industries in Japan (an intraindustry restructuring), rather than predominantly replacing them (an interindustry restructuring). An important element here is that in this phase Japanese investments in dynamic industries take place mainly in other industrialized countries with established tastes and technology in the same sectors. Competitive interaction with these American and European demand-and-supply potentials often provides creative challenges to the Japanese subsidiaries and may generate new ideas that prove useful throughout the group's operations. These potentials are a central element of the behavior we are concerned with in this chapter.[7]

The view that overseas operations in this third phase of Japanese FDI not only facilitate the continued and extended use of established capabilities (as in phases one and two) but also encompass the potential to help to regenerate and upgrade them is endorsed by Ozawa, who notes (1991a, 137) that "the rich cultural diversities and creative human resources in Europe are the new resources Japan needs—and can tap only through direct overseas operations—in its search for ideas and knowledge." Dunning also suggests that the Japanese overseas investments of the latter part of the 1980s represent a move from defensive strategies to a more offensive approach where "such strategies have been largely driven by the need of Japanese firms to transform themselves from exporters to 'insiders' in the major markets of the world, and to

6. The change of direction was enshrined in a 1971 policy directive issued by the Industrial Structure Deliberation Council, a body acting as adviser to the Ministry of International Trade and Industry (MITI).

7. Ozawa (1991a, 143–44; 1991b, 57; 1992, 17–18) discerns the emergence during the 1980s of a new (fourth) phase of industrialization in Japan, based around computer-aided flexible manufacturing. This phase is likely to eventually provoke a fourth wave of overseas investments exhibiting radically new characteristics, which, however, can only be speculative at this point (but see Ozawa 1991b, 57–60).

keep in touch with the latest technological and organizational developments" (Dunning 1993a, 149).

Data provided by JETRO on the cumulative overseas investments of Japanese companies in manufacturing between 1950 and 1992 shows that electrical machinery is now clearly the most strongly established industry in the sector, with 23.5 percent of this cumulative investment total. As table 1.4 shows, mainly third-phase investments have now established Europe and North America as major recipients of Japanese FDI in this industry. Asia also retains a notable position in Japanese electrical companies' FDI, which certainly involves some operations that have persisted from first-phase investments but also includes additional third-phase investments to play an important supply role in the new global strategies of these enterprises as well as to supply the increasingly significant host-country markets in the region.[8] With 14.0 percent of cumulative manufacturing FDI covered by the JETRO data, chemicals is marginally the second-largest manufacturing industry. Asia again stakes a strong position that probably developed out of the "housecleaning" (pollution exporting) second phase of Japanese FDI, while the significant shares now established in North America and Europe surely include an important element of knowledge-seeking third-phase operations, deriving from pharmaceuticals, biotechnology, and other innovative parts of the industry. The JETRO data shows that transport machinery claims 13.5 percent of the cumulative manufacturing FDI, with the strongest focus being third-phase investments in North America and Europe. An essentially resource-seeking second-phase industry, ferrous and nonferrous metals, claims the next-largest share of the cumulative investment (11.6 percent), being of notably above-average importance in investments in Latin America and also in Asia. The last of the mainly third-phase industries, machinery, accounted for 9.9 percent of the cumulative FDI, again revealing the expected focus on developing-country operations.[9]

## Japanese MNEs in Europe

The previous part of this section has traced the considerable strengthening of the position of Japanese companies in the global stocks and flows of FDI and has also noted the position of investors in Europe in this overall picture. In

8. MITI data analyzed by the United Nations Centre on Transnational Corporations (UNCTAD 1991, figure V, 49), shows that in 1987, 22.0 percent of the sales of Japanese companies' electrical and electronic equipment subsidiaries in Asia were exported to Japan, 19.3 percent to other countries outside Asia, and 15.7 percent to other countries within the Asia region.

9. Of the remaining industries, food accounted for 5.0 percent of cumulative FDI, textiles for 4.8 percent, and timber and pulp for 3.6 percent.

**TABLE 1.4. Distribution of Japanese Manufacturing Direct Investment Abroad by Industry, Cumulative Value, 1950–92 (in percentage)**

| | Europe | North America | Latin America | Asia | Oceania | Total[a] |
|---|---|---|---|---|---|---|
| Food | 11.4 | 49.4 | 5.1 | 26.7 | 7.2 | 100.0 |
| Textiles | 22.4 | 21.0 | 9.6 | 45.8 | 0.3 | 100.0 |
| Lumber and pulp | 3.3 | 67.6 | 6.5 | 16.5 | 6.1 | 100.0 |
| Chemicals | 13.8 | 40.7 | 6.0 | 36.3 | 0.9 | 100.0 |
| Ferrous and nonferrous metal | 6.1 | 41.8 | 17.8 | 27.5 | 5.3 | 100.0 |
| Machinery | 27.7 | 45.8 | 4.7 | 20.5 | 1.2 | 100.0 |
| Electrical machinery | 21.5 | 51.9 | 3.1 | 22.8 | 0.5 | 100.0 |
| Transport machinery | 21.4 | 44.9 | 10.5 | 14.7 | 8.5 | 100.0 |
| Other | 11.2 | 65.5 | 1.2 | 21.4 | 0.9 | 100.0 |
| Total manufacturing | 16.7 | 48.4 | 6.6 | 23.7 | 2.8 | 100.0 |

*Source:* JETRO (mimeo distribution).
*Note:* Calculations are based on approval basis up to 1980 and on reported basis since then.
[a]Includes Africa and Middle East.

this part we look in more detail at aspects of the operations of Japanese manufacturing MNEs in Europe. Table 1.5 shows that the number of affiliates of Japanese MNEs in manufacturing industries in Europe increased by more than five times from 1983 to 1993. A persistent growth through the decade established the United Kingdom as clearly the leading host country by 1993, with France also doubling its share to emerge into second place, ahead of

**TABLE 1.5. Japanese Manufacturers Operating in Europe, by Country, 1983–93**

| | Number of Production Bases[a] | | | | Share (%) | | | |
|---|---|---|---|---|---|---|---|---|
| | 1983 | 1987 | 1990 | 1993 | 1983 | 1987 | 1990 | 1993 |
| United Kingdom | 15 | 53 | 129 | 198 | 12.8 | 21.9 | 25.0 | 27.8 |
| France | 10 | 33 | 89 | 121 | 8.5 | 13.6 | 17.3 | 17.0 |
| Germany | 20 | 45 | 86 | 107 | 17.1 | 18.6 | 16.7 | 15.0 |
| Netherlands | 13 | 20 | 33 | 46 | 11.1 | 8.3 | 6.4 | 6.5 |
| Belgium and Luxembourg | 13 | 18 | 27 | 42 | 11.1 | 7.4 | 5.2 | 5.9 |
| Ireland | 9 | 10 | 22 | 30 | 7.7 | 4.1 | 4.3 | 4.2 |
| Spain | 18 | 29 | 54 | 63 | 15.4 | 12.0 | 10.5 | 8.8 |
| Italy | 7 | 11 | 28 | 45 | 6.0 | 4.5 | 5.4 | 6.3 |
| Austria | 2 | 5 | 12 | 17 | 1.7 | 2.1 | 2.3 | 2.4 |
| Portugal | 4 | 6 | 13 | 13 | 3.4 | 2.5 | 2.5 | 1.8 |
| Other | 6 | 12 | 22 | 31 | 5.1 | 5.0 | 4.3 | 4.3 |
| Total | 117 | 242 | 515 | 713 | 100.0 | 100.0 | 100.0 | 100.0 |

*Source:* JETRO 1993, table I.1, 3.
[a]Numbers at end of January of each year.

**TABLE 1.6. Japanese Manufacturers Operating in Europe at end of January 1993, by Country and Industry**

| | United Kingdom | | France | | Germany | | Benelux | | Total | |
|---|---|---|---|---|---|---|---|---|---|---|
| | No. | % | No. | % | No. | % | No. | % | No. | % |
| Food | 3 | 1.5 | 20 | 16.5 | 3 | 2.8 | 5 | 5.7 | 32 | 4.5 |
| Textiles and apparel[a] | 7 | 3.5 | 5 | 4.1 | 1 | 0.9 | 0 | 0 | 30 | 4.2 |
| Chemicals and pharmaceuticals | 31 | 15.7 | 22 | 18.2 | 16 | 15.0 | 28 | 31.8 | 136 | 19.1 |
| Metals[b] | 17 | 8.6 | 6 | 5.0 | 8 | 7.5 | 4 | 4.5 | 56 | 7.9 |
| General machinery | 15 | 7.6 | 11 | 9.1 | 19 | 17.8 | 8 | 9.1 | 73 | 10.2 |
| Electronic and electrical equipment | 41 | 20.7 | 22 | 18.2 | 19 | 17.8 | 10 | 11.4 | 112 | 15.7 |
| Electronic components | 24 | 12.1 | 9 | 7.4 | 18 | 16.8 | 7 | 8.0 | 72 | 10.1 |
| Transport equipment and parts | 26 | 13.1 | 8 | 6.6 | 3 | 2.8 | 6 | 6.8 | 72 | 10.1 |
| Precision instruments | 9 | 4.5 | 4 | 3.3 | 10 | 9.3 | 5 | 5.7 | 35 | 4.9 |
| Others | 25 | 12.6 | 14 | 11.6 | 10 | 9.3 | 15 | 17.0 | 95 | 13.3 |
| Total | 198 | 100.0 | 121 | 100.0 | 107 | 100.0 | 88 | 100.0 | 713 | 100.0 |

*Source:* JETRO 1993, table I.2, 4.

[a]Textiles, apparel, and textile products.

[b]Iron and steel, nonferrous metals, and metal products.

Germany. Spain and the Benelux countries suffered the main losses in share over the period. Certain areas of host-country specialization, in terms of industries attracted, can be discerned in table 1.6. Notable among these are the relative strengths of France in food, Benelux in chemicals and pharmaceuticals, Germany in general machinery, Germany and the United Kingdom in electronics (including components), and Germany in precision instruments.

Data compiled by Dunning (table 1.7) makes it possible to compare the industrial patterns (as reflected in employment levels) of Japanese and U.S. investments in Europe. The industries most strongly associated with the third phase of Japanese FDI, electronics and transport equipment, have established the two leading positions among the total of Japanese MNEs' manufacturing employment in Europe, at levels somewhat more prominent than those of the same two industries in U.S. affiliates' employment. The very prominent position taken by Japanese affiliates in the rubber industry reflects acquisitions (notably that of Dunlop by Sumitomo) that may have perhaps now established adequate capacity to support associated car makers operating locally, so that further growth of investment in the industry may not be expected to match that of the lead industries or to reflect typical patterns of Japanese competitive

**TABLE 1.7. Employment Distribution of Japanese and U.S. Affiliates in European Manufacturing**

| | Japan[a] Employment | | USA[b] Employment | | |
|---|---|---|---|---|---|
| | (000s) | (%) | (000s) | (%) | Ratio[c] |
| *Processing industries* | | | | | |
| Food and kindred products | 453 | 0.50 | 119.7 | 7.95 | 0.06 |
| Pharmaceuticals | 1,823 | 2.01 | 79.5 | 5.28 | 0.38 |
| Other chemical products[d] | 4,800 | 5.29 | 177.7 | 11.81 | 0.45 |
| Iron and steel[e] | 450 | 0.49 | 36.5 | 2.43 | 0.20 |
| Textiles and clothing | 493 | 0.54 | 30.6 | 2.03 | 0.27 |
| Wood-related products | 14 | 0.02 | 6.9 | 0.46 | 0.04 |
| Rubber products | 18,778 | 20.72 | 40.8 | 2.71 | 7.65 |
| Nonferrous materials[f] | 4,010 | 4.42 | 72.5 | 4.82 | 0.92 |
| *Fabricating industries* | | | | | |
| General machinery | 8,384 | 9.25 | 158.5 | 10.53 | 0.88 |
| Electronic and electrical equipment[g] | 22,981 | 25.36 | 268.6 | 17.85 | 1.42 |
| Transport equipment | 21,061 | 23.24 | 309.1 | 20.54 | 11.10 |
| Instruments | 3,905 | 4.31 | 107.2 | 7.12 | 0.61 |
| Other industries | 3,459 | 3.82 | 97.5 | 6.48 | 0.59 |
| Total manufacturing | 90,611 | 100.00 | 1,505.1 | 100.00 | 1.00 |

*Source:* Dunning 1994, 62.
[a]1990.
[b]1989.
[c]Share of industry in all Japanese affiliates' employment divided by share of same industry in all U.S. affiliates' employment.
[d]Includes plastics.
[e]Includes metal products.
[f]Includes glass, stone, and ceramics.
[g]Includes office machinery and computers.

advantage. Chemicals and pharmaceuticals have now reached modestly significant positions in overall Japanese affiliate employment in Europe but still clearly take less prominent positions than among U.S. affiliates.

Valuable comparative data on the supply of the European market by Japanese and U.S. firms have been compiled by Dunning and Cantwell (1991, 160–62). These estimates[10] make it possible to compare not only the relative levels of sales in Europe by firms from these two countries, but also their composition in terms of two alternative modes of supply, namely exports to Europe from the firm's home country and production carried out in owned and controlled operations in Europe.[11] Overall, Japanese firms' sales in Europe

10. Though the authors admit that the compilation of the comparative data sets involved some "heroic" assumptions (Dunning and Cantwell 1991, 183), there is little doubt that the evidence with regard to broad patterns and differences is convincing and significant.

11. A third way of supplying the European market would be through licensed production

rose from $9.8 billion in 1976 to $38.2 billion in 1986, while those of U.S. firms increased from $112.2 billion to $243.0 billion. Thus the Japanese firms' share of the total European sales of these two groups of firms rose from 8.1 percent in 1976 to 13.6 percent in 1986. In fact, to some extent, the rise in Japanese firms' share of the European market occurred through both of the alternative routes of supply. Thus Japanese enterprises accounted for 39.6 percent of the exports to Europe by the two groups of companies in 1976 and for 42.7 percent in 1986, while their share of production in Europe by affiliates increased from 1.2 percent to 3.4 percent.

The figures just quoted indicate the key difference between the European market orientation of Japanese and U.S. firms, with production in Europe being very strongly established in U.S. enterprises but with exports to Europe retaining the leading position in Japanese companies. Nevertheless the share of the sales in Europe of Japanese firms accounted for by their affiliates' production there had risen from 12.3 percent in 1976 to 18.5 percent in 1986, while the comparable figures for U.S. firms indicated a small decline from 88.2 percent to 82.8 percent. As Dunning and Cantwell point out (1991, 161), "these figures alone suggest that there is a very considerable potential for more Japanese production in Europe." It is, however, a theme of the arguments to be developed here that the expansion of Japanese enterprises' output in Europe that has occurred since 1986, and which seems likely to continue to some degree throughout the 1990s, does not merely represent the substitution of local production for goods that would otherwise have been imported from other sources in these companies. The behavior discussed in later chapters indicates that Japanese MNEs' subsidiaries in Europe acknowledge that fully effective investments must seek to be market enhancing, by using their increased local knowledge and through access to additional sources of creative assets there, to develop new products (or at least substantially modified established ones) that more completely satisfy consumers' needs.

For 1986 Dunning and Cantwell (1991, table 9.5) are able to extend their estimates to cover individual manufacturing industries. Once again the two industries in which Japanese firms were most distinctively well established in the European market in 1986 were electrical equipment and transport equipment, these accounting for 37.2 percent and 26.8 percent, respectively, of overall Japanese enterprises' sales in Europe, for 34.3 percent and 27.1 percent of their production there, and for 37.9 percent and 26.7 percent of their

carried out by European firms using some productive asset (e.g., technology) supplied by a Japanese or U.S. firm. Dunning and Cantwell (1991, table 9.4) provide estimates for such nonaffiliate licensing sales. Thus in 1986, sales under license from Japanese firms in Europe were estimated to be 16.5 percent of those deriving from exports plus affiliate production, an increase from 12.4 percent in 1976. The comparable ratio for U.S. firms was 5.5 percent in 1986 and 4.6 percent in 1976.

exports to those markets. By contrast, electronic equipment only accounted for 10.1 percent of U.S. firms' overall European sales, for 4.8 percent of local production, and for 35.6 percent of exports, while the comparable shares for U.S. transport equipment enterprises were 16.0 percent, 16.7 percent, and 12.4 percent, respectively. Three industries in which Japanese companies' European production was relatively weak by comparison with that of U.S. firms were machinery (24.8 percent of U.S. European production compared with 9.4 percent of Japanese), chemicals (21.0 percent compared with 7.4 percent), and food, drink, and tobacco (14.4 percent compared with 3.1 percent). Since U.S. firms in these industries were also rather the stronger in terms of exporting to Europe, these sectors also accounted for notably smaller shares of total Japanese companies' sales in Europe (10.2 percent, 7.7 percent, and 0.9 percent for machinery, chemicals, and food, drink, and tobacco, respectively, compared with 23.0 percent, 20.0 percent, and 12.9 percent shares of U.S. firms' sales).

Some perspectives on the motivation of Japanese manufacturing operations in Europe can be gained from a JETRO survey in 1992, which specified 16 possible reasons for setting up such production facilities. Respondents were permitted to endorse any of these that were perceived as relevant. Table 1.8 includes the 10 motives that were supported by more than 10 percent of the 386 respondents. Broadly, it is clear that the most pervasive motives were those relating to the European market and conditions that indicated that production within that market was increasingly the most effective way of supplying it. Thus the most often-supported motive was the general need "to meet expanding demand." However, the decision to respond to this expanding demand by production in Europe emerges as deriving from a mixture of positive and negative factors relating to market conditions.

Among the more negative demand-related factors providing motivation for Japanese firms to establish manufacturing operations in Europe, "to meet a rise in trade protectionism resulting from the EC's market integration" takes the most prominent position. Two other motives relating to potential trade problems also emerge among the 10 most significant as reported in table 1.8; namely, "to avoid discriminatory quantitative restrictions on Japanese imports" and "to avoid infringement of antidumping regulations." Each of these three motives was perceived as being of well-above-average significance for firms operating in the processing and assembly sector. Two factors that are also reflections of possible trade distortions are included among the six less influential motives not reported in the table. Thus, "to avoid exchange rate fluctuations" was endorsed by 6.7 percent of respondents and "to avoid infringement of antidumping regulations on parts and components" by 5.4 percent.

Other demand-related motives may be seen as more optimistic, as they

**TABLE 1.8. Motives for Establishment of Japanese Manufacturing Operations in Europe**

| Sector | No. of replies | A | B | C | D | E | F | G | H | I | J |
|---|---|---|---|---|---|---|---|---|---|---|---|
| Processing and assembly[a] | 127 | 46.4 | 29.0 | 45.7 | 29.9 | 11.8 | 11.8 | 44.1 | 22.8 | 6.3 | 6.3 |
| Parts and components[b] | 100 | 48.0 | 15.0 | 37.0 | 33.0 | 8.0 | 18.0 | 33.0 | 9.0 | 5.0 | 36.0 |
| Raw materials[c] | 146 | 42.5 | 6.8 | 26.0 | 32.2 | 11.6 | 16.4 | 34.2 | 4.1 | 6.8 | 17.8 |
| Other | 34 | 52.9 | 11.8 | 14.7 | 20.6 | 5.9 | 14.7 | 50.0 | 8.8 | 5.9 | 8.8 |
| Design and/or R&D centers | 23 | 17.4 | 8.7 | 17.4 | 21.7 | 17.4 | — | 17.4 | — | 82.6 | 4.3 |
| Total[d] | 386 | 44.0 | 15.5 | 33.7 | 31.1 | 10.6 | 14.2 | 36.5 | 10.4 | 10.6 | 16.3 |

*Source:* JETRO 1992b, table III-I, 17.

*Note:* Motives were defined as follows:

A. to meet expanding demand.
B. to avoid discriminatory quantitative restrictions on Japanese imports.
C. to meet a rise in trade protectionism resulting from the European Community's market integration.
D. to benefit from the expansion of economic activity resulting from the European Community's market integration.
E. strategy in anticipation of expanded European markets resulting from German unification, the realization of the European Economic Area initiative, as well as from the liberalization of East European markets.
F. investment incentives such as tax credits.
G. to meet consumer's needs.
H. to avoid infringement of antidumping regulations.
I. to carry out R&D activities in Europe.
J. to supply parts, components, and raw materials to affiliates of Japanese manufacturers operating in Europe.

Numbers in each motive column reflect the percentage of replying firms that considered the motive relevant.

[a]General machinery; electronic and electronic appliances; transport equipment; precision machinery.

[b]Electronic components and transport equipment parts.

[c]Manufacturing and/or production of: foodstuffs; textiles, clothing, and made-up goods; furniture; pulp and paper; chemicals; pharmaceuticals; rubber products; ceramics, stone, and clay products; iron and steel; nonferrous metals and metal products.

[d]Some respondents were included in two industries, so the total is less than the sum of the sectors.

respond to a more positive view of European markets or represent the desire to react to them in a more committed manner than is normally possible through trade. Thus the second most popular motive, "to meet consumers' needs," may reflect a growing perception of the commercial benefits to be gained from seeing European consumers as representing a distinctive and important segment of the global market, and one that can best be served through local production. The establishment of Japanese manufacturing operations in Europe was also quite widely seen as aiming "to benefit from the expansion of economic activity resulting from the EC's market integration." Other potential sources of wider markets in the region had not yet emerged as a very significant stimulus to Japanese firms, however, with relatively few considering their investments as part of a "strategy in anticipation of expanded European markets resulting from German unification, the realization of the European Economic Area initiative, as well as from the liberalization of East European markets."

Two of the factors investigated in the JETRO survey may be classified as derived-demand motives, in that they represent a response to an earlier investment decision of another firm. Thus table 1.8 shows that a moderate number of the subsidiaries that replied were established "to supply parts, components, and raw materials to affiliates of Japanese manufacturers operating in Europe." In addition 9.8 percent of respondents considered they were "treading on the heels of parent companies that have already advanced in Europe."

Turning to what may be considered to be supply-side factors, three of the motives included in the survey relate to influences that could boost profitability by lowering costs and improving the ability to supply the firm's markets. Of these, only "investment incentives such as tax credits" attained a position in the 10 most important motives, while "to reduce production costs" was endorsed by only 8.8 percent of respondents and "to secure raw materials at lower cost" by only 3.4 percent. It should be noted, however, that while these factors clearly have limited relevance to the frontline decision to set up operations in Europe, as investigated in the survey, they may take a stronger position in the secondary decision as to where in the region a facility should be located.

Finally, two of the motives related to the performance of more creative activities in the subsidiary. As would be expected, table 1.8 shows that the vast majority (82.6 percent) of design and/or R&D centers indicated that "to carry out R&D activities in Europe" was a motive in their establishment. Equally predictably, few of the subsidiaries in the four manufacturing industries (overall 6.1 percent) considered this a motive behind their establishment, though many of them clearly possess such facilities to support their operations. Overall only 6.5 percent of respondents endorsed the motive "to implement design and development operations in Europe" (not included in table), with this motive once again being considerably more prevalent among design and/or R&D centers (34.8 percent) than in production subsidiaries (4.6 percent).

## The Rise of Japanese Technology

In this section we provide evidence on aspects of the rise of technology in the operations of Japanese enterprises. The first part illustrates the growth of R&D intensity (R&D expenditures as a proportion of sales) as an indicator of increased inputs into technology creation, then provides evidence on Japanese firms' patenting as an output measure. The second part looks at the involvement of Japanese companies in strategic technology alliances, another means of expanding research and technological scope.

## Japan's R&D and Patenting Performance

Using data from the 1991 Economic White Paper of the Economic Planning Agency, Yoshitomi (1992) demonstrates the widespread rise of Japanese industry's commitment to R&D. The data covers not only the direct R&D support of an industry (i.e., the R&D of its own firms) but also the indirect support embodied in improved intermediate and investment goods available to it as a result of R&D carried out in input-supplying industries and calculated by using input-output tables. Table 1.9 shows that all sectors analyzed benefited from an overall increase in R&D support (direct and indirect), with the strongest benefit in the electrical, transport, and precision machinery industries and in chemicals. Yoshitomi notes that electrical and precision machinery industries were the ones that were particularly reliant on strong growth in their own R&D, while transport machinery and iron and steel benefited most from indirect technological advances.

Further clear evidence on the increasing importance to Japanese companies of technological creativity and innovation-oriented activities can be derived from patenting in the United States[12] by the world's leading enterprises.[13] Since these leading international firms will usually seek to protect their ability to compete globally, it is felt that their U.S. patenting is likely to provide an internally consistent and comprehensive comparative reflection of their technological performance. The data on each patent granted in the United States records both the location of the research facility originally responsible for the innovation and the firm to which the patent has been granted. Work carried out in collaboration between researchers at the University of Reading and the Science Policy Research Unit of the University of Sussex made it possible to establish the ultimate ownership of patents where they had been granted to affiliates of MNEs. This meant that for a large number of the world's leading industrial enterprises, data was available on the total number of patents granted to the group in the United States in a particular year and on the number of these that were attributable to research performed

12. The data on the geographical origins of patents granted in the United States have been compiled at the University of Reading in collaboration with the Science Policy Research Unit at the University of Sussex. The data on U.S. patent counts was prepared by the Office of Technology Assessment and Forecast, U.S. Patent and Trademark Office, with the support of the Science Indicators Unit, U.S. National Science Foundation. The opinions expressed here are those of the authors and do not necessarily reflect the view of the Patent and Trademark Office or the National Science Foundation.

13. The data covers all the firms from the sample of the 792 largest industrial enterprises in the world in 1982, analyzed by Dunning and Pearce (1985, 8–11), for which patenting records were available, plus 41 additional companies excluded from that sample but where patenting was extensive.

**TABLE 1.9. Total R&D Intensity of Japanese Industry, Direct and Indirect, 1975–89 (in percentage)**

| | 1975 | 1985 | 1989 |
|---|---|---|---|
| Agriculture | 0.60 | 0.71 | 0.79 |
| Mining | 0.71 | 0.71 | 1.40 |
| Construction | 0.81 | 1.04 | 1.31 |
| Foods | 0.83 | 1.04 | 1.32 |
| Textiles | 0.96 | 1.43 | 1.83 |
| Paper and pulp | 0.89 | 1.31 | 1.60 |
| Chemicals | 1.93 | 5.34 | 7.13 |
| Petroleum and coal | 0.77 | 0.83 | 1.67 |
| Ceramics | 1.20 | 2.00 | 2.69 |
| Iron and steel | 1.49 | 3.69 | 4.36 |
| Nonferrous metals | 1.58[a] | 2.79 | 2.66 |
| Metals | 1.11 | 2.19 | 2.26 |
| General machinery | 1.85 | 3.12 | 3.82 |
| Electrical machinery | 2.16 | 6.43 | 7.57 |
| Transport machinery | 1.62 | 5.23 | 6.13 |
| Precision machinery | 1.51 | 5.34 | 5.91 |

*Source:* Yoshitomi 1992.
[a]1980 figure.

outside the MNE's home country. In chapter 3 we use this data to assess the position of overseas operations in the creative activity of Japanese enterprises, while in chapter 4 it helps us evaluate certain aspects of such activity in their subsidiaries in Europe. Meanwhile the patent data reported in tables 1.10 and 1.11 can provide us with indicators of the overall growth of technological operations in leading Japanese enterprises.

It is clear that table 1.10 reveals Japanese companies as having had a uniquely distinctive commitment to increased technological creativity during the period 1969 to 1990. Thus the Japanese firms increased their share of total patents from 6.1 percent in 1969–72 to 30.3 percent in 1987–90, which took them from an initial position that was comparable to that of the United Kingdom or Germany into a clearly decisive second place in the patenting league, at a level that had reached two-thirds of that of U.S. firms, compared with less than one-tenth at the start of the period. The Japanese firms in fact took out 498 percent more patents in the United States in 1987–90 than they did in 1969–72, compared with a rise of only 20 percent for all leading enterprises covered by the data. By contrast, the levels of U.S. patenting of U.S. and U.K. firms in 1987–90 were 24 percent and 18 percent, respectively, *lower* than they had been in 1969–72.

The increasing commitment of Japanese enterprises to technology as an important facet of their approach to competitiveness is also revealed, in table

**TABLE 1.10. Shares of Total Patents Granted to the World's Largest Firms in the United States, by Home Country of Firm (in percentage)**

| | 1969–72 | 1973–77 | 1978–82 | 1983–86 | 1987–90 |
|---|---|---|---|---|---|
| Japan | 6.1 | 10.9 | 15.9 | 23.2 | 30.3 |
| United States | 71.4 | 62.9 | 57.5 | 52.0 | 45.1 |
| Europe (total) | 21.9 | 25.5 | 25.8 | 24.0 | 23.6 |
| European Community (total) | 18.3 | 20.7 | 21.3 | 20.1 | 20.2 |
| United Kingdom | 6.4 | 5.6 | 4.9 | 4.3 | 4.3 |
| Germany | 7.4 | 9.7 | 10.8 | 10.2 | 10.2 |
| France | 2.1 | 2.6 | 2.6 | 2.6 | 2.6 |
| Netherlands | 1.7 | 2.0 | 1.9 | 2.0 | 2.4 |
| Other European (total) | 3.6 | 4.7 | 4.5 | 3.8 | 3.4 |
| Sweden | 1.0 | 1.2 | 1.1 | 1.1 | 0.9 |
| Switzerland | 2.5 | 3.3 | 3.1 | 2.4 | 2.1 |
| Total | 100.0 | 100.0 | 100.0 | 100.0 | 100.0 |

*Source:* U.S. patent data base held at the University of Reading, with the support of the Patent and Trademark Office, U.S. Department of Commerce.

1.11, to be widely dispersed by industry. Thus in the 12 industries covered separately in the table, Japanese firms had accounted for more than 10 percent of 1969–72 patenting in only one case, professional and scientific instruments. By contrast, in 1987–90 Japanese firms accounted for *less* than 10 percent of patenting in only two industries (pharmaceuticals; petroleum and coal), while they claimed over half the patents in two (instruments; motor

**TABLE 1.11. Japanese Companies' Shares of Total Patents Granted to the World's Largest Firms in the United States, by Industry (in percentage)**

| | 1969–72 | 1973–77 | 1978–82 | 1983–86 | 1987–90 |
|---|---|---|---|---|---|
| Food, drink, and tobacco | 3.7 | 6.2 | 7.1 | 8.2 | 10.9 |
| Chemicals | 6.2 | 10.3 | 12.5 | 17.6 | 22.8 |
| Pharmaceuticals | 4.0 | 4.2 | 4.6 | 5.0 | 5.2 |
| Metals | 6.1 | 10.9 | 14.2 | 22.2 | 30.9 |
| Mechanical engineering | 1.1 | 3.2 | 6.3 | 8.3 | 12.4 |
| Electrical engineering | 9.7 | 15.9 | 23.0 | 33.7 | 42.3 |
| Office equipment | 5.3 | 6.6 | 10.7 | 18.0 | 25.0 |
| Motor vehicles | 9.7 | 21.1 | 34.3 | 46.3 | 54.2 |
| Rubber and plastic products | 2.6 | 9.4 | 14.2 | 18.8 | 29.5 |
| Nonmetallic minerals | 6.9 | 17.0 | 20.7 | 25.9 | 29.8 |
| Petroleum and coal | 0.2 | 0.8 | 0.9 | 1.4 | 2.9 |
| Professional and scientific instruments | 14.3 | 34.7 | 48.9 | 60.5 | 60.6 |
| Other | 4.5 | 9.2 | 10.9 | 11.2 | 9.9 |
| Total | 6.1 | 10.9 | 15.9 | 23.2 | 30.3 |

*Source:* U.S. patent data base at the University of Reading, with the support of the Patent and Trademark Office, U.S. Department of Commerce.

vehicles) and a quarter or more in five others (electrical engineering; metals; nonmetallic minerals; rubber and plastic products; office equipment). The Japanese companies increased their share of patenting in all of the 12 industries, with this increase being greater than 10 percentage points in all except three (food, drink, and tobacco; pharmaceuticals; petroleum and coal) and exceeding 30 percentage points in three (electrical engineering; motor vehicles; instruments).

### Strategic Technology Alliances

Many of the pressures that induce firms to diversify their technological scope and options by increasingly decentralizing their own R&D have also resulted in an increasing tendency toward technology-motivated strategic alliances between otherwise independent firms.[14] Many of these strategic technology alliances (STAs) have themselves been international (or in our discussion "interbloc," treating Europe as an entity alongside the United States and Japan). The extent and nature of Japanese firms' involvement in such STAs can be analyzed using data on 4,192 such collaborations established between 1980 and 1989 as compiled in the Cooperative Agreements and Technology Indicators (CATI) data bank, compiled at the Maastricht Economic Research Institute on Innovation and Technology (MERIT) of the University of Limburg.[15] For purposes of the analysis of the MERIT-CATI data base, STAs have been defined (Hagedoorn 1993a, 372) as "interfirm cooperation for which a combined innovative activity or an exchange of technology is at least part of their agreement. The strategic character of the agreement relates to its expected long-term effects on the product-market positioning of at least one of the partners."[16]

Within the overall context of STAs, as just defined, the MERIT-CATI data is divided into two broad categories, according to the predominant perceived motivation for the collaboration. First, R&D-driven alliances are seen as responding to "motives related to basic and applied research and some general characteristics of technological development" (Hagedoorn 1993a, table 1, 373). In addition to the reduction and sharing of the costs and uncer-

14. For conceptual and theoretical background to the emergence of STAs, see Chesnais 1988, 78–88; Hagedoorn and Schakenraad 1991, 7–25; Dunning 1993a, 194–204; and Cantwell 1993, 24–25.

15. For details on the compilation of this data base, see Hagedoorn 1993a, 377–78; and Hagedoorn and Schakenraad 1993a, 78–9. For general discussion of possible limitations on the reliability of this type of data base, see Chesnais 1988, 64–65. For possible problems relating explicitly to coverage of alliances involving Japanese companies, see Chesnais 1988, 68–70.

16. For detailed discussion of the organizational modes of STAs, see Hagedoorn 1993b, 127–34.

tainty of R&D, these alliances are also perceived as a reaction to the "increased complexity and intersectoral nature of new technologies, cross-fertilization of scientific disciplines and fields of technology, monitoring of evolution of technologies, technological synergies, access to scientific knowledge or to complementary technology." In the analysis of the MERIT-CATI STAs, two particular motives that can be encompassed within the R&D-driven context emerged as particularly important. Thus 31 percent of the 4,192 STAs were to some degree motivated by "technological complementarity" and 28 percent by the pursuit of "reduction of innovation time span" (Hagedoorn 1993a, table 3, 379). Therefore this type of cooperation "has to be understood in the light of attempts made by companies to cope with the complexity and interrelatedness of different fields of technology and their efforts to gain time and reduce uncertainty in joint undertakings during a period of growing technological intricacy" (Hagedoorn 1993a, 378). As will be seen in the next chapter, our arguments relating to the emergence of global-innovation strategies in MNEs are set very much in this environment, as a response to these technological developments and competitive forces.

The second broad group of STAs distinguished in the MERIT-CATI data analysis is seen as related to market access and the search for opportunities. These STAs therefore involve a "combination of market access and technology development through a combined effort of companies . . . cooperation is found in the opportunities for market entry through a joint monitoring of environmental changes in combination with new products or processes" (Hagedoorn 1993a, 374). It could be that a more distinctive division of labor may occur in this type of STA, with perhaps one partner possessing strong technological capabilities but feeling relatively weak in marketing aspects, while the other has well-developed marketing capacity but considers its technology vulnerable. This configuration might then be especially relevant in an interbloc context, with a technologically powerful firm feeling insecure in the extension of its operations into another market due to skepticism about the scope of its marketing expertise. An STA with a local firm in that market (perhaps one that feels a complementary weakness in its own technology) may then provide inputs that not only improve the firm's ability to sell its current products there, but also contribute to appropriate product and process adaptations that further enhance competitiveness. The fact that generally table 1.13 reveals R&D-driven STAs to be less prevalent in interbloc than intrabloc STAs is compatible with this interpretation of the second motivation.[17]

17. Hagedoorn and Schakenraad (1993, 74) express this result as "R & D-driven alliances tend to be less international and still closer to the home bases of companies," interpreting it in terms of traditional arguments for persisting centralization of R&D, "economies of scale and economies of scope for research activities can limit the spread of such facilities" (72).

Overall the MERIT-CATI data on STAs demonstrates that Japanese firms are, in absolute terms, certainly increasing use of this approach to the creation and application of technology, but also that it still seems relatively less strongly ingrained in their behavior than is the case for comparable U.S. or European companies. Thus table 1.12 shows that in 1985–89 Japanese enterprises were involved in 622 of the STAs recorded in the MERIT-CATI data base (adding totals for Europe-Japan, Japan-United States, and Japan), a rise of 41.4 percent over the 440 of 1980–84. However, the comparable rises for U.S. and European firms were 60.8 percent and 76.2 percent, respectively. The figures of table 1.12 also show that in 1985–89 Japanese companies were involved in 23.6 percent of the recorded STAs, compared with 28.2 percent in 1980–84. For European firms the comparable figures were 48.3 percent and 57.4 percent for the two periods, and for U.S. firms, 59.6 percent and 62.6 percent.[18] The sector in which Japanese firms have established the strongest presence in STAs is new materials, where they were involved in 43.8 percent of the total in 1985–89, compared with 30.6 percent in 1980–84. By contrast, Japanese companies lost position substantially in information technology STAs, being involved in only 20.6 percent of the total in 1985–89, compared with 34.2 percent in 1980–84, and also, to a lesser degree, in other technologies (22.1 percent compared with 25.6 percent). Their involvement in biotechnology STAs remained at a low level, with their participation at 20.3 percent in the latter period, representing a marginal rise from the 19.2 percent of the former.

The results reported in table 1.13 indicate that there was a distinct tendency during the 1980s for STAs to increasingly support the efforts of firms in the creation of new technology. Thus R&D-driven STAs generally revealed a quite widely pervasive propensity to increase their share of the total, with, for example, significant rises in three of the four industrial sectors. Japanese companies participated quite extensively in this increased orientation to R&D-driven STAs. Thus it can be calculated from the MERIT-CATI data that 57.4 percent of the STAs involving Japanese enterprises in 1985–89 were assessed as being R&D-driven, compared with 49.5 percent in 1980–84. The comparable figures for STAs involving European firms were very similar to those for Japanese (59.7 percent and 50.6 percent), but the move was rather stronger for U.S. companies (up to 68.3 percent from 53.5 percent). The notable surge in Japanese companies' involvement in STAs in new materials

18. Thus Japanese, European, and U.S. firms all record participation in smaller shares of total STAs in 1985–89 than in 1980–84. This is numerically possible because the interbloc STAs (e.g., between Japan and Europe), which are incorporated twice (i.e., in the calculation for both Japanese and European firms) in the calculations, were relatively less important in 1985–89 than in 1980–84 compared with the intrabloc STAs (e.g., between Japanese firms), which are included only once.

**TABLE 1.12. Distribution of Strategic Technology Alliances within and between Economic Blocs, 1980–89**

| | Biotechnology | | New Materials | | Information Technology | | Other Technology | | Total | |
|---|---|---|---|---|---|---|---|---|---|---|
| | 1980–84 | 1985–89 | 1980–84 | 1985–89 | 1980–84 | 1985–89 | 1980–84 | 1985–89 | 1980–84 | 1985–89 |
| Europe-Japan | 5 | 20 | 15 | 23 | 57 | 57 | 24 | 50 | 101 | 150 |
| Japan-United States | 45 | 54 | 16 | 40 | 133 | 132 | 80 | 83 | 274 | 309 |
| Japan | 11 | 33 | 7 | 71 | 29 | 35 | 18 | 24 | 65 | 163 |
| Europe | 47 | 95 | 29 | 44 | 85 | 217 | 115 | 173 | 276 | 529 |
| Europe-United States | 58 | 124 | 32 | 52 | 158 | 256 | 97 | 161 | 345 | 593 |
| United States | 125 | 179 | 16 | 54 | 142 | 323 | 74 | 111 | 357 | 667 |
| Other | 27 | 23 | 9 | 22 | 37 | 68 | 69 | 108 | 142 | 221 |
| Total | 318 | 528 | 124 | 306 | 641 | 1,088 | 477 | 710 | 1,560 | 2.632 |

*Source:* Hagedoorn and Schakenraad 1991, 69–73.

**TABLE 1.13. Share of R&D-driven STAs in total STAs, by Area and Industry, 1980–89 (in percentage)**

| | Biotechnology | | New Materials | | Information Technology | | Other Technologies | | Total | |
|---|---|---|---|---|---|---|---|---|---|---|
| | 1980–84 | 1985–89 | 1980–84 | 1985–89 | 1980–84 | 1985–89 | 1980–84 | 1985–89 | 1980–84 | 1985–89 |
| Europe-Japan | 60.0 | 80.0 | 46.7 | 47.8 | 50.9 | 45.6 | 20.8 | 38.0 | 43.6 | 48.0 |
| Japan-United States | 82.2 | 87.0 | 25.0 | 60.0 | 46.6 | 55.3 | 35.0 | 20.5 | 47.8 | 52.1 |
| Japan | 81.8 | 90.9 | 85.7 | 94.4 | 72.4 | 51.4 | 44.4 | 37.5 | 67.7 | 76.1 |
| Europe | 68.1 | 81.1 | 51.7 | 75.0 | 52.9 | 58.1 | 45.2 | 37.0 | 52.2 | 56.7 |
| Europe-United States | 75.9 | 87.9 | 40.6 | 71.2 | 43.7 | 55.5 | 52.6 | 61.5 | 51.3 | 65.3 |
| United States | 64.8 | 88.3 | 62.6 | 79.6 | 63.4 | 75.2 | 44.6 | 71.2 | 59.9 | 78.4 |
| Other | 59.3 | 82.6 | 33.3 | 40.9 | 21.6 | 44.1 | 24.6 | 17.6 | 31.0 | 34.8 |
| Total | 69.8 | 86.4 | 46.8 | 73.2 | 50.5 | 60.5 | 40.7 | 43.1 | 51.2 | 62.5 |

*Source:* Hagedoorn and Schakenraad 1991, 69–73.

noted earlier emerges in these additional calculations as having apparently been particularly influenced by R&D needs. Thus 76.1 percent of the new materials STAs in which Japanese firms participated in 1985–89 were recorded as R&D-driven compared with 44.6 percent of those in 1980–84. However, the MERIT-CATI data shows that biotechnology remained the sector where Japanese firms (like those from Europe and the United States) were most likely to use STAs for R&D-motivated purposes, with 86.9 percent of those in 1985–89 (compared with 80.3 percent in 1980–84) assessed as having this orientation. In information technology the orientation of STAs involving Japanese companies was more balanced, with those considered as R&D-driven being 51.1 percent of the total in the first period and 52.2 percent in the second. Finally, the STAs in which Japanese enterprises participated in other technologies had a persistently below-average propensity to be R&D driven, with those considered to have this orientation falling from 33.6 percent to 28.7 percent.

CHAPTER 2

# Internationalized Research and Development

In this book the emergence of internationalized R&D in Japanese, and in longer-established, MNEs is seen as a key component in new decentralized strategies for the creation and application of technology among such globally competing enterprises. The adoption of these new approaches can in turn be perceived as the logical reaction to significant changes in the worldwide competitive environment. This reflects the long-standing ability of the most successful companies to find modes of operation that enable them to achieve effective aggressive or defensive responses to new potentials or problems that emerge in their producing and marketing situations. Useful background to the new perspectives in MNE behavior that are investigated here can be derived from Dunning's (1993b, 56–63) typology of overseas production. This typology distinguishes four types of MNE activity, though it is noted that the contemporary MNE often has pluralistic objectives and engages in several of the types of operations.

## Types of Multinationals

### Natural Resource Seekers

The pursuit of various types of resources has always been a motivation for FDI and continues to play a central role in the contemporary strategies we analyze here. In Dunning's typology, natural resource seekers are "enterprises [that] are prompted to invest abroad to acquire particular and specific resources at a lower real cost than could be obtained in their home country" (1993b, 57). Three main types of resource seekers can be discerned.

First, there are those that seek physical resources such as minerals, raw materials, and agricultural products. Primary producers and manufacturing enterprises have traditionally engaged in this type of FDI, motivated by the pursuit of cost minimization and security of supply of vital inputs. Dunning observes that on the eve of World War II approximately three-fifths of the accumulated FDI stake was in this kind of operation, but that by the mid-1980s its share had fallen to around one-third. The second type of resource-seeking MNE has sought "plentiful supplies of cheap and well-motivated unskilled or semi-skilled labor" (1993b, 57). Thus manufacturing- and service-sector MNEs "from countries with high real labor costs" use facilities "in countries with lower real labor costs, to supply labor intensive intermediate or final products for export" (57). However, Dunning suggests that a declining role for these types of labor in the value-adding process in several manufacturing industries has reduced the scope for this type of resource seeking.

The third, and most contemporary, type of resource-seeking activity is the one of greatest relevance to our work. Thus Dunning discerns FDI that is "prompted by the need of firms to acquire technological capability, management or marketing expertise and organizational skills" (1993b, 57). Whereas the previous two resource-seeking roles may have passed their respective peaks, our analysis supports Dunning's view that "FDI to gain access to technology, information and specialized managerial skills is more important than it used to be" (58). As the sources of these key knowledge assets become more globally dispersed and differentiated, firms pursuing competitiveness in dynamic industries based on such intangible inputs need to seek the optimal supply mix of them worldwide. As the pressures of ever changing market competition and technological needs at the firm level combine with the evolving ability of different countries to supply various types of knowledge assets, sensitivity in this third type of resource seeking becomes central to the type of competitive behavior we discuss here.

### Market Seekers

In Dunning's typology, market-seeking MNEs "invest in a particular country or region to supply goods or services to markets in these or in adjacent countries" (1993b, 58). The extension of this definition to regional, rather than purely national, markets is especially relevant to our discussion as it is clear[1] that the market-seeking element in Japanese companies' investments in Europe usually covers markets beyond that of their immediate host country. This broadened definition reflects the widening context and motivation of market-seeking FDI, which "may be undertaken to sustain or protect existing markets or to exploit or promote new markets." Thus in addition to characteristics of markets in terms of their size, growth, and income levels,[2] Dunning discerns five motives for market-seeking investments.

The first, and in Dunning's view still "the single most important reason for market-seeking investment" (1993b, 59), is action by host-country gov-

1. See Papanastassiou and Pearce 1993, 1994c.

2. For the influence of such factors on U.S. and U.K. FDI, see Papanastassiou and Pearce 1994b, 1990. Pearce (1993, 81–82) surveys earlier analyses of market influences on the determinants of FDI.

ernments. Historically this approach took the form of import-substituting operations, in which new host-country tariffs or other trade restraints made it desirable for successful firms to replace exports with overseas production in order to continue to supply their existing markets. The second, and analytically quite similar, motive suggests that when a foreign market becomes sufficiently large, the production of goods that are relatively expensive to transport, and that do not benefit greatly from economies of scale in production, may be relocated within the market itself (or to a more adjacent cost-effective site).

The third motive is that component-supplying firms may implement overseas production in order to retain the market of their established customers when these have already set up producing operations abroad. The notable relevance of this motive to the growth of Japanese FDI in Europe is well known and has been observed in the discussion in the previous chapter (e.g., see table 1.8). A fourth motive in market-seeking investment, which Dunning assesses as of increasing relevance, is "that a MNE may consider it necessary, as part of its global production and marketing strategy, to have a physical presence in the leading markets served by its competitors" (1993b, 58–59). The importance of key markets as distinctive elements in the battleground of global competition, as seen by leading players in oligopolistic industries, is central to the conditions determining much of the behavior described in our subsequent analysis.

However, it is the fifth of Dunning's market-seeking roles in FDI that provides the most direct and immediate motivation for the extension in MNEs' technological operations that we discuss, namely that "quite frequently products need to be adapted to local tastes or needs, and to indigenous resources and capabilities" (1993b, 58). In fact, this much more positive and market-expanding type of FDI can be perceived as having quite interesting theoretical precedents. In an influential discussion of international trade, Linder (1961) suggested[3] that in two developed countries A and B at similar levels of per capita income, there is likely to be a substantial area of common or overlapping taste for the products of a particular industry, but that there are also likely to be areas of distinctive taste in each country that reflect the more idiosyncratic elements in each country's characteristics. Responding to the areas of common taste, the more successful firms in country A can export at least some of their established products successfully to country B, and vice versa. In this manner intraindustry trade can emerge in quite narrowly defined product-differentiating industries. However, if an already competitive firm from country A decides to set up production facilities in country B, then the increased knowledge it can thereby obtain of market, production, and techno-

3. See Ozawa 1992, 19–23.

logical conditions in that country may enable it to use local resources to adapt existing products or develop new products. These new products then extend its market beyond the area of overlapping taste, which it could supply by exports, into the area reflecting those more distinctive local taste characteristics that it would have been very much harder to respond to effectively through trade. Of course similarly competitive firms from country B have the same market-expanding opportunity from setting up production operations in country A. In effect the FDI behavior of these firms extends the area of overlapping taste beyond that which would have been available through trade, with potentially welfare-enhancing consequences. It also provides an explanation of intraindustry FDI that parallels that of intraindustry trade. Caves' (1971) view that the ability to differentiate products is a significant factor in facilitating the international spread of MNEs' operations is also clearly compatible with this line of argument.

### Efficiency Seekers

As international trade has become freer and global competition has intensified, it has become increasingly both necessary and feasible for MNEs to use each of their subsidiaries in the way that can most effectively support the overall worldwide competitiveness of the group. Thus Dunning observes the emergence of efficiency-seeking MNEs that aim "to rationalize the structure of established resource-based or market-seeking investment in such a way that the company can gain from the common governance of geographically dispersed activities" (1993b, 59). The intention of this rationalization is "to take advantage of different factor endowments, cultures, institutional arrangements, economic systems and policies, and market structures by concentrating production in a limited number of locations to supply multiple markets" (59). In this way each individual subsidiary in an efficiency-seeking MNE aims to realize to the full the benefits from a source of competitive advantage that it can derive from particular characteristics of its host country.[4] The group then seeks to assemble a balanced and coordinated network of such specialized subsidiaries that will allow it to maximize its competitiveness in its worldwide markets.

### Strategic Asset Seekers

Finally Dunning notes a fourth group of MNEs which comprise "those which engage in FDI, usually by acquiring the assets of foreign corporations, to

4. Though simple economies of scale, which do not particularly reflect distinctive host-country characteristics, are sometimes the basis of such a subsidiary.

promote their long-term strategic objectives—especially that of sustaining or advancing their international competitiveness" (1993b, 60). This development clearly relates strongly to the increased emphasis on the geographical dispersion of technological competences in MNEs that is the subject of this chapter. Thus an MNE may take over a foreign company to obtain its existing technology and its established research capacity. Alternatively an MNE may create its own R&D facility abroad, either to be an integral part of a production and marketing subsidiary or to operate independently of such local linkages, in order to tap into a high-quality national science base that can thereby enhance and complement relevant areas of its existing technology. Furthermore the strategic technology alliances discussed in the previous chapter may be seen as primarily a manifestation of strategic-asset-seeking behavior.

In moving toward the pursuit of such strategic-asset-seeking behavior, it can be suggested that Japanese companies first became adept at using external sources of technology during the era of licensing and reverse engineering. This ability persisted into the era when Japanese firms were becoming more independently innovative (Mansfield 1988). However, whereas reverse engineering involved more effective use of embodied technology through product and process *adaptation,* the subsequent phase involved increasing pursuit of new disembodied technology that could be *developed* in distinctive ways. Westney and Sakakibara (1985, 316–17) emphasize the ability that emerged in Japanese companies to use their overseas production and marketing operations as a part of a global technology scanning and monitoring effort. As a part of this process, Japanese enterprises discovered the relevance to their innovation and long-term science programs of centers of technological capacity in the United States, Europe, and elsewhere.[5] This discovery stimulated the location of R&D units in these areas for two possible reasons. First, it may have been decided that it is better to internalize parts of these pools of capability into their own operations, rather than rely on the partial and shared access to them that could be secured through external relationships.[6] Second, when it is decided that it is not desirable or reasonable to internalize particular elements of such scientific capacity, it is most effective to articulate the externalized relationship with them through local R&D laboratories. Such laboratories may do extensive complementary research themselves or may serve mainly as an organizing or clearinghouse center for a network of external research links. In general terms, a dual role for R&D has been analyzed by Cohen and Levinthal (1989, 593), who argue "that firms invest in R&D not only to pursue directly new process and product innovation, but also to

5. The way in which Japanese MNEs have sought new and complementary technology through their operations in the United States has been analyzed by Kogut and Chang (1991).

6. For discussion of this decision, see Cantwell 1993, 12–18.

develop and maintain their broader capabilities to assimilate and exploit externally available information."

## Types of MNE Subsidiaries

### New Environment for Subsidiaries

As the scope and aims of globally competing firms have evolved and widened, the nature and position of individual subsidiaries within such MNE groups have also undergone important changes. These subsidiary-level developments are crucial in influencing the emergence of significant decentralized technological activity in MNEs and in determining the forms it can take. The background to the emergence of new roles for overseas subsidiaries lies in changes in the international competitive environment and in the position of MNEs within it.

A significant element in these changes has been the progressive lowering of tariff protection through the series of General Agreement on Tariffs and Trade (GATT) negotiation rounds. This improvement in the conditions for international trade removes much of the motivation and support for the traditional import-substituting types of market-seeking FDI. Though tariffs have often been replaced by the more ad hoc trade barriers of the new protectionism, it can be argued[7] that these rarely promise the level of stability that would provide reliable support for new, or continued, local-market-focused import-substituting market-seeking investments. The trade environment that incorporates the new protectionism as one element requires new strategic approaches by MNEs. For instance, though we have observed (in the first chapter) that actual or anticipated protection has stimulated much of Japanese manufacturing investment in Europe, it is also clear that these operations usually develop an efficiency-seeking mode of behavior, using the freedom of trade within the European Union (the European Community before November 1993) to implement a network of rationalized subsidiaries in order to optimize the supply of products throughout the integrated regional market.

Indeed, the emergence of Japanese firms as major MNEs, followed increasingly by additional competitors from the newly industrializing countries, represents a second significant change in the environment of global competition. The ability of these firms to bring to their international production the assets that had already made them formidably competitive through international trade, and their willingness to implement their overseas production through strategies that immediately emphasize efficiency-seeking approaches, force the established traditional MNEs to more fully recognize the pressures

7. See Pearce 1992, 40.

on their often inefficient long-standing market-seeking production facilities and to also pursue new strategic approaches to restore the globally competitive viability of their operations.

A third new element adding to the intensity of global competition and influencing the strategic approaches of old and new MNEs is the increased international dispersion of technological capability. Through the 1950s and into the early 1960s, U.S. and European MNEs had something approaching a monopoly of competitive technology, often derived from the commercial application of knowledge created for military needs in World War II and the early years of the cold war. Also, during those years, these established MNEs were the most systematically oriented to the use of R&D facilities and scientific personnel in order to create new technology and enhance the competitive scope of their existing technology. This orientation, then, constituted a major ownership advantage that enabled them to sustain competitiveness in many markets, even though their market-seeking strategies did not always permit or require individual production facilities to operate in an optimally efficient manner. However, during the 1960s, and increasingly in more recent years, Japanese and other firms first of all became increasingly effective users of existing technology, then subsequently implemented the independent capability to allow them to develop their own distinctive technology. Thus a dominance of technology quickly receded as a unique source of competitive strength that could compensate for other inefficiencies in the operations of the longer-established MNEs.

Recovery from this new vulnerability has involved two main responses in the behavior of these MNEs: (1) the restructuring of their production operations to employ more thoroughly and systematically the potentials of subsidiaries that carry out specialized roles that embody inputs that reflect particular asset strengths of their host country, and (2) revitalization of their technology, and the renewal of its position in their strategy, by adopting a global approach to the acquisition and application of key elements of knowledge. This behavior brings technology and knowledge into strategic focus among the resource-seeking priorities of FDI and also underpins the emergence of increasingly strategic-asset-seeking modes of organization. The growing momentum behind these developments in global competition is central to the analysis of this chapter and to the behavior of Japanese enterprises in Europe discussed in subsequent chapters. The remaining parts of this section define three types of subsidiaries[8] whose characteristics embody the strengths and weaknesses of traditional and emergent MNE strategies.

---

8. For similar approaches to the classification of subsidiaries, see, for example, Crookell and Caliendo 1980, 61–62; White and Poynter 1984, 59–61; Young, Hood, and Hamill 1987, 155–58; Cordell 1971, 29–30; and Crookell 1986, 104–7.

### Truncated Miniature Replica Subsidiaries

A truncated miniature replica (TMR) subsidiary adopts a classic form of market-seeking behavior, supplying a substantial part of its MNE group's established product range to its host-country market, usually with the protection of tariffs or other trade barriers. These TMRs are thus import-substituting subsidiaries that are part of horizontally integrated MNEs that adopt what Porter (1986) has termed a multidomestic strategy. A TMR is thus considered, at least in terms of its production and marketing activity, to function as a miniature replica of the parent and other leading subsidiaries in the MNE group. However, it is also perceived (especially by critical host-country governments) as being truncated, in the sense that its operations omit certain activities that would normally be considered an essential part of a healthy independent business. For the purposes of this exposition we can emphasize the technological dependence of TMRs and the very limited role of R&D in them. Thus TMRs may incorporate adaptation work to make the product or its production process more suitable for the local environment but would be unlikely to embody the scope or motivation for any more fundamentally innovative work.

Clearly it is TMR subsidiaries whose inefficiencies (small-scale local-market-focused production, lack of independent creativity)[9] are most likely to be exposed by the changes in the international competitive environment set out in the previous section. However, since MNEs usually wish to retain the markets of those host countries previously supplied by these TMRs, they often consider that simply closing such facilities (in the hope of continuing to supply the market through imports) risks high levels of unpopularity that would be potentially severely damaging to their competitive position. Thus these market-seeking subsidiaries have often evolved into taking on new roles that are more strongly based around other types of MNE motivation. The typology discerns two such possibilities.

### Rationalized Product Subsidiaries

One possible reorientation of subsidiary roles has been toward rationalized operations, in which individual subsidiaries take up specialized positions in a broadly defined global or regional strategy. An individual rationalized product (RP) subsidiary might, for example, produce only a very limited part of the MNE's product line. What it produces will be determined by its capabilities, rather than by the particular needs of its host-country market. The most relevant capabilities will be those that predominantly reflect distinctive aspects of the host-country producing conditions, since the subsidiary is ex-

9. See Pearce 1992, 39–42.

pected to have equal access to all firm-level capabilities (technology, know-how, management, and so on) so that its capacities are unlikely to differ greatly from other RP subsidiaries in those respects. This indicates that within a predominant efficiency-seeking motivation, RP subsidiaries also embody strong elements of resource seeking, with cost-effective labor and other physical inputs key priorities.

It is usually intended that much of an RP subsidiary's output will serve foreign markets, while the parts of the local-market demand that it does not meet are supplied from elsewhere in the MNE group. The RP subsidiary's strong local producing presence will be expected to create the goodwill that makes such imports acceptable to local consumers, in a manner that might not have been possible if such imports had replaced the output of a TMR that had been closed down.

When these subsidiaries can rationalize their output, compared to the attempt to produce a wide product range in a TMR, this specialization permits a much greater level of efficiency through the realization of economies of scale. An obvious variation on this theme involves RP subsidiaries specializing in the production of particular component parts for assembly into final goods elsewhere or, alternatively, performing a separate stage in a vertically integrated production process. In all these cases the aim is to achieve efficient production by making optimum use of the distinctive productive capabilities of different locations accessible to the MNE.

By performing a specialized role at optimal efficiency, an RP subsidiary escapes from the deficiencies of a miniature replica. It is likely, however, to remain severely functionally truncated. It will have little scope for independence in management or marketing and will be extremely constrained with respect to independent technological creativity. Since an individual RP subsidiary produces an established product for external markets, it seems clear that it will not undertake product adaptation. Some process adaptation is a possibility, but the likelihood of this being necessary is limited by the fact that the subsidiary should be allocated a production role to match its existing capacities. However, some R&D may be necessary where a network of RP subsidiaries supplies a regional market with a range of products first innovated elsewhere (e.g., where Japanese or U.S. MNEs supply European markets with products created in their home countries). In such a case, one R&D facility could serve the whole R&D network and, for reasons of group politics, may be located independently of any producing facility.

### Product Mandate Subsidiaries

A central theme of this book is that the overseas operations in MNEs often have the potential to provide a wider range of inputs than those merely involved in efficient production of established products. In some locations at least, management, marketing, and technological capabilities are available well beyond those needed by RP subsidiary operations. Therefore we can distinguish a second possible direction of evolution for outmoded TMRs into world or regional product mandate (WPM or RPM) subsidiaries, which make an integrated creative use of a wider range of talented local inputs. WPMs have been defined by Bonin and Perron (1986, 161) as "an agreement between a multinational enterprise's parent company and one of its subsidiaries to grant the subsidiary exclusive rights to produce and market a product and, if circumstances warrant, to pursue the necessary research and development activity." The mandated subsidiary thus becomes the international (or regional) center for a product, with responsibility for all the key inputs relating to its creation, production, and marketing.

The efficiency-seeking motivation of WPM or RPM operations thus extends beyond the production efficiency of the RP subsidiary to embody a more dynamic creative efficiency, in a manner that also extends the complementary resource-seeking objectives. Thus, while efficient physical inputs are still relevant to production operations (as is central to RP subsidiaries), the wider range of value-adding activity in product mandate subsidiaries means that their resource-seeking priorities focus most strongly on technological capability, along with managerial, marketing, and organizational expertise. In addition strategic-asset-seeking behavior in MNEs may complement the presence of WPM or RPM operations.

Many of the documented cases of WPMs[10] emerged in an ad hoc fashion, with an entrepreneurial subsidiary detecting an opportunity (sometimes an idea underdeveloped elsewhere in the group) and securing permission to bring it to fruition. In some of these cases the subsidiary's own technological resources can adequately achieve the product's full development. Alternatively the WPM can, of its own volition, secure complementary scientific inputs from elsewhere in the group, while retaining overall responsibility for product development.[11] A less ad hoc role, which may become of particular relevance to RPMs, would provide them with key positions in a global approach to innovation. As will be discussed in more detail shortly, we can conceive of a broad new product idea emerging at headquarters level (though often as a result of the operations of decentralized basic and applied research facilities), with responsibility for the quick and responsive commercialization of the product in different major regions devolving to local RPM subsidiaries. Thus these subsidiaries are able to combine their local development, engineering, and marketing knowledge and ability to define a product variant that is fully attuned to local needs and conditions.

10. See Pearce 1992, 48–51 for illustrative examples.

11. See Pearce 1989, 122–30 for more detailed analysis of R&D in WPMs.

### A Global-Innovation Strategy

Alongside the need to find more efficient or creative roles for their overseas subsidiaries, a second contemporary priority for MNEs is an enhanced commitment to the generation of new technology and its effective commercial application. The argument of this section suggests that ultimately these two behavioral developments come together in the logic of a globalized approach to innovation, which will incorporate key roles for geographically decentralized R&D and for the more creative types of overseas subsidiary.

Interesting evidence on the emerging perceptions of new technology-competitive pressures in globally operating enterprises can be derived from the results of a survey and questionnaire analysis by Granstrand and Sjolander, summarized in table 2.1.[12] Though the respondents to the survey and questionnaire were specifically technology executives, it would seem likely that they would be sufficiently senior to have a broadly based view of company priorities, so that table 2.1 may be considered as generally indicating the substantial relative significance of technology-based issues in contemporary decision making in Japanese and U.S. enterprises.[13]

Within the 12 technology issues reported in table 2.1, "demand for higher quality" and "keeping pace with new product technologies" reflect the continued crucial importance of sensing and responding to trends in market needs and technological developments in established product areas. However, the increasing significance of more radical product creation activities in contemporary competition can be discerned in the strong Japanese and U.S. firm responses to "shorter market lifetime of products," "pressure for more frequent introduction of new generations of products," and "pressure for shorter innovation lead times" as vital issues. Japanese respondents provide relatively strong valuations of "increased complexity (fusion) of technology" and "increased fusion between science and technologies," indicating that these firms are either leading or finding problems with a trend toward a widening technological basis for operations in many leading industries. Though Japanese respondents did not provide a strong response to "pressure to acquire technology from abroad," increased internationalization of R&D operations would be

12. The Granstrand and Sjolander study "is based on 91 interviews with technology executives among 12 Swedish, 14 Japanese and 16 U.S. corporations and on questionnaire surveys on the corporate and product area level, respectively, among these firms. All of the firms are technology-based, multinational and among the largest in each country" (Granstrand and Sjolander 1992, 183).

13. Granstrand and Sjolander (1992, 186) suggest that a possible explanation for the lower priority for technology-related issues in Swedish respondents is "that Swedish companies are much more niche-oriented, and often dominate or are among the dominating firms within their niches, and have a better control over their immediate commercial environment."

**TABLE 2.1. Importance of Managerial Issues and Country Differences as Perceived by Technology Executives in Multitechnology Corporations in the United States, Japan, and Sweden**

| | United States | Japan | Sweden |
|---|---|---|---|
| Keeping pace with new product technologies | 2.75 | 2.82 | 2.33 |
| Escalating R&D spending | 2.12 | 2.27 | 1.17 |
| Pressure for shorter innovation lead times | 2.88 | 2.54 | 1.83 |
| Shorter market lifetime of products | 2.62 | 2.64 | 1.67 |
| Pressure for more frequent introduction of new generations of products | 2.50 | 2.45 | 2.00 |
| Pressure to acquire technology from outside the company | 2.12 | 2.09 | 1.17 |
| Pressure for technological protectionism | 1.88 | 1.82 | 0.83 |
| Pressure for scientific protectionism | 1.50 | 1.64 | 0.67 |
| Pressure to acquire technology from abroad | 2.00 | 1.73 | 0.67 |
| Increased complexity (fusion) of technology | 2.50 | 2.70 | 1.67 |
| Increased fusion between science and technologies | 2.00 | 2.45 | 1.00 |
| Demand for higher quality | 2.88 | 2.82 | 3.00 |
| Average for 12 technology issues[a] | 2.31 | 2.33 | 1.50 |
| Average for 16 nontechnology issues[b] | 1.74 | 1.76 | 1.59 |

*Source:* Granstrand and Sjolander 1992, table 9.1, 185.

*Note:* The following scale was used. 0 = unimportant, 1 = of minor importance, 2 = important, 3 = of major importance.

[a]Unweighted average of individual responses for the 12 issues.

[b]Unweighted average of individual responses for the 16 non-technology issues.

a logical response to this need to integrate a wider range of scientific competences, especially in the light of a perceived weakness of Japan's capability in certain key areas of technology.

Granstrand and Sjolander (1992, 186–87) also report that "a trend analysis in managerial perceptions between two time periods, 1983–87 and 1988–92, reveals expectations for a more challenging environment globally as perceived by managers in the three countries." Issues that were reported as significantly more important in the second period by executives in the three countries were shorter innovation lead times and market lifetimes, increased external acquisition (sourcing) of technology, technological protectionism, increased fusion of technologies and of science with technology, and increased demand for higher quality.

A fully developed global-innovation strategy will embody three objectives.[14] First, it will pursue the cost-effective performance of the basic research that is necessary to enable the company to move forward its underlying

14. See Papanastassiou and Pearce 1994c, 776–77.

technological base in a manner that provides for its sustained long-term commercial evolution. Though the evidence of table 2.1 suggests that the costs of research are certainly likely to be of some concern in this objective, the main target will be to obtain high-quality inputs in all the areas of science that can make a contribution to these precompetitive research programs.

The second objective of the global-innovation strategy is to innovate new products as quickly as possible in *all key markets* internationally. As results reported in table 2.1 again make clear, the needs of contemporary competition would not be met adequately by a process of centralized innovation, as envisaged by Vernon's (1966) original product cycle, followed by the new products only gradually entering into foreign markets, first through exports and only later through overseas production.[15] The more aggressive response to overseas markets within the process of innovation, which is implied by this second objective of a global-innovation strategy, takes on a further dimension in the third. It suggests that in many globally competitive industries, the areas of distinctive and idiosyncratic tastes in key markets, which were noted in the discussion of market-seeking behavior in an earlier section, may expand faster than the areas of overlapping taste as incomes rise. If growing incomes allow consumers to manifest their national taste characteristics more strongly in this way, MNEs that can respond to such preferences within the innovation process itself can gain a vital competitive edge. Thus the third objective of a global-innovation strategy is to get new products into separate foreign markets, not only quickly, but in differentiated forms that represent full acknowledgment of their distinctive characteristics and needs. In this way the global approach to innovation provides a greatly extended scope to the product-differentiation element that has existed in earlier forms of market-seeking behavior, moving, in effect, from product adaptation to product development.

The objectives of a global-innovation strategy may be achieved through a two-phase approach. The aim of the first phase is to derive the outlines of a new product concept, which extends the scope of the industry in quite radical ways. At this stage the characteristics of the new concept fully define what the product does, and the way in which it does it, without providing complete details of the precise form in which it will be offered to consumers or the way in which it is to be produced. In this first phase the MNE will use all the basic and applied research rcsources available to it to establish the new technology that underpins the innovative product concept. The work of this stage is likely to be carried out by a global network of pure and precompetitive-research-

15. Vernon (1979) later reviewed the precepts of the product cycle in light of subsequent developments. Bartlett and Ghoshal (1989, 1990) provide an important conceptualization of emerging approaches to global innovation that also reflect new needs of the competitive environment.

oriented laboratories, which can access the areas of technology where specialist inputs are needed. Thus, as Cantwell's (1991) research has shown, the range of technologies needed by firms in many technologically dynamic industries has widened, while the areas of outstanding technological capacity of many countries have narrowed. In this situation it is logical for science-based MNEs to pursue an optimal selection of research inputs to their precompetitive research by tapping into the specialist capabilities of several countries. The network of such laboratories is likely to be strongly coordinated by a "parent" laboratory, probably in the MNE's home country, that takes responsibility for the ultimate formulation of the new product concept.

If the first phase seeks to achieve the first objective of a global-innovation strategy, the second phase deals with the other two objectives. Thus a second set of product-development laboratories, often operating as part of RPM subsidiaries, takes up the outlines of the new product concept and proceeds to define its detailed characteristics and the precise form of the production process in order to meet the specific needs of their own markets and production conditions. In this second phase the new product reaches global markets through a series of more-or-less simultaneous differentiated innovations that result in a set of product variants, each of which meets the detailed needs and conditions of a separate national or regional market.

## Roles of Overseas R&D Facilities

The scenario of a full-scale global-innovation strategy outlined in the previous section clearly implies both more extensive and more diversified roles for overseas R&D laboratories in MNEs. We now introduce a typology of overseas laboratory types[16] that takes account of the emerging roles for such facilities.

### Support Laboratories

Support laboratories (SLs) assist an MNE's production and marketing facilities overseas to make effective use of the group's existing technology. Thus SLs deal with problems experienced in implementing established products and processes efficiently in new locations, often some considerable time after the original innovation has been completed. Two particular possibilities for this role have been observed.[17] First, the SL may act as a technical service

16. For earlier typologies of overseas laboratories, see Ronstadt 1977, 1978; Hood and Young 1982; Haug, Hood, and Young 1983; Hakanson 1981; Cordell 1971, 1973; and Pearce 1989.

17. See Cordell 1971, 46–49; 1973, 23–25.

center, by examining why a product may fail to operate in a local market or by helping with the adaptation of the product for the local market. Second, the SL may further assist the process of technology transfer by advising on any necessary adaptation of the production techniques used, which could involve the need to adjust production techniques for shorter runs and to scale down engineering designed for longer runs than the host-country market can support.

Thus SLs could support TMRs in making the best use of existing technology in their markets, adapting either products or processes as needed. SLs could also play roles in relation to RP subsidiaries. When supporting one such RP unit directly, an SL's work is likely to be limited to process adaptation, as adaptations of the product to specific host-country conditions should be irrelevant in a predominantly export-oriented subsidiary. However, if an SL supports a complete network of RP subsidiaries, it may then do work on the products, in light of the larger market supplied, as well as on production processes.

### Locally Integrated Laboratories

Locally integrated laboratories (LILs) involve themselves in much more clearly development-oriented activity than the problem-solving SLs. By working closely and creatively with other functions in the same subsidiary (i.e., marketing, engineering, management), LILs seek to determine all the details of a product that they develop in light of the characteristics of their target market and production environments.

An LIL may emerge by superseding an SL in a particular country when an ambitious TMR subsidiary management perceives its local market to be sufficiently distinctive and fast growing to merit the development of products that are much more than marginal adaptations of parts of its MNE's existing range. Perhaps by gaining permission to address the technology underlying current products at the stage of applied-research results, an LIL could, in a sense, redevelop the product by deriving a substantially new variant within the existing group technology. Such LIL operations would seem likely to focus on a very limited number of products. However, the more successful of these may be sufficiently radical to create new regional or global markets outside the host country, which could then lead to the subsidiary evolving into a WPM or RPM, as may well have been the ambition of its managers.

In the context of the more systematic approaches to global innovation discussed in the preceding section, a more formal role for LILs within RPM operations clearly emerges. Thus it would be the LIL that established a close collaboration with central R&D in order to familiarize itself with the technology behind the new product concept that has been derived from phase one of the global-innovation strategy. It would then take this knowledge back to its RPM subsidiary in order to collaborate closely with marketing, engineering, and various parts of management to derive the variant of the concept that is most likely to be competitively effective in its target market area. In this way an LIL is a key agent in fulfilling the second phase of the global-innovation strategy.

### Internationally Interdependent Laboratories

The internationally interdependent laboratory (IIL), though located in a particular country, has its main orientation toward the precompetitive part of the MNE's global R&D activity, which aims to supply mainly basic and applied work to an internationally coordinated program that seeks to provide the technological basis for a new generation of products. The IIL is therefore clearly located within phase one of the global-innovation strategy. Because of this motivation, it has no obvious reason to be systematically linked with any other operations of the MNE in the same country, which would be more concerned with the effective local implementation of already commercialized technology. While not integrated locally, such a laboratory is interdependent with similar laboratories in other countries, which play complementary roles in the wider phase one program. Then individual responsibilities within a global network of IILs will be allocated according to the specialized capabilities that each can derive from the areas of particular expertise that have emerged within national science bases.

## Extent and Characteristics of Overseas R&D

A survey carried out at the University of Reading in 1989–90 provides evidence on the distribution and roles of overseas R&D laboratories in MNEs (Pearce and Singh 1992). This survey used two separate questionnaires to investigate the internationalization of R&D in 560 of the world's leading enterprises.[18] The first questionnaire in this survey was sent to the parent laboratories of the companies, these being either corporate-level R&D units (either physically located at corporate headquarters or otherwise distinguished as the corporate unit) or the main R&D facilities of major divisions in diversified enterprises. Several of the 560 companies were indeed diversified, with

18. The 560 enterprises comprised the 500 largest industrial enterprises in the world in 1986, as derived from the *Fortune* listings, plus 30 more for which directories of R&D facilities revealed to have substantial numbers of overseas laboratories, plus 30 similar companies without overseas R&D (to match the 30 self-selecting overseas-R&D-oriented enterprises). For more details, see Pearce and Singh 1992, 9–16.

their main divisions having their own parent laboratories, so that 623 were surveyed in total. Usable replies were ultimately obtained from 245 of these parent units. The other questionnaire in this survey was sent to the 405 identified overseas subsidiary laboratories of the 560 companies. These are nonparent units located outside the home country of the group. Satisfactory replies were secured from 133 of these subsidiary laboratories.

The parent laboratory questionnaire provides some broad indications of the current distribution of overseas laboratories in MNEs. A question in the survey asked responding parent laboratories to list all their subsidiary R&D units by country. Overall 211 of the parent labs provided this information, covering a total of 1,315 separate facilities, of which 785 were in the MNE groups' home countries and 530 were overseas. The U.S. respondents reported an average of 3.75 laboratories at home and 1.72 overseas, or 0.45 labs abroad for each one at home. Responding Japanese MNE parent laboratories reported an average of 5.44 units in Japan and 4.00 overseas, an average of 0.74 labs abroad for each at home. For United Kingdom-based respondents there was an average of 2.71 units at home and 2.26 abroad, or 0.85 laboratories overseas for each one at home. This pattern is stronger for Continental European MNEs, which reported 4.54 laboratories abroad per parent compared with 3.92 at home, or 1.18 abroad for each one in the home country. The strongest influence on this result, as is indicated in table 2.2, is their dispersion of R&D facilities *within* the Other Europe area.

In table 2.2 we look in more detail at the distribution of the overseas R&D laboratories reported by responding parent units. In the columns denoted A we simply provide the numbers of overseas labs for each MNE parent country, by industry and host country. The columns denoted B then provide an indication of the intensity of commitment to overseas R&D by expressing the numbers of R&D facilities as a proportion of the number of relevant responding parent laboratories; for example, for U.S. respondents in pharmaceuticals and consumer chemicals there were 3.62 labs overseas per parent lab, while all U.S. parents averaged 0.38 labs in the United Kingdom. This perspective confirms the strong commitment to internationalized R&D by continental European MNEs, which recorded 4.54 labs abroad per responding parent unit. Though, as noted previously, the key factor here is an extensive dispersion of R&D facilities within the area, those Other European MNEs also incorporate a broader perspective on globalized R&D, usually matching the typical levels of commitment to other areas and countries as well. The industries where these continental European MNEs are most distinctively oriented to R&D abroad are chemicals and pharmaceuticals.

The notably high numbers of R&D units reported per Japanese parent laboratory owes much to the electronics and pharmaceutical industries and is especially reflected through operations in the United States. Apart from a

**TABLE 2.2. Overseas R&D Laboratories of Leading MNEs, by Parent Country, Host Country, and Industry**

| | Parent Country of MNE | | | | | | | | | |
|---|---|---|---|---|---|---|---|---|---|---|
| | United States | | United Kingdom | | Other Europe[a] | | Japan | | Total[b] | |
| | A | B | A | B | A | B | A | B | A | B |
| *By industry* | | | | | | | | | | |
| Food, drink, and tobacco | 0 | 0 | 13 | 1.30 | 11 | 3.67 | — | — | 29 | 1.26 |
| Industrial and agricultural chemicals | 31 | 1.72 | 9 | 2.25 | 67 | 13.40 | 5 | 1.67 | 112 | 3.73 |
| Pharmaceuticals and consumer chemicals | 29 | 3.62 | 11 | 2.75 | 25 | 12.50 | 8 | 8.00 | 73 | 4.87 |
| Electronics and electrical appliances | 59 | 4.92 | 6 | 3.00 | 33 | 5.50 | 69 | 11.50 | 167 | 6.42 |
| Industrial and farm equipment | 10 | 1.67 | — | — | 15 | 3.00 | 0 | 0 | 25 | 2.09 |
| Motor vehicles | — | — | 3 | 1.50 | 9 | 3.00 | 14 | 2.80 | 26 | 2.60 |
| Metal manufacture and products | 9 | 1.13 | 13 | 2.17 | 5 | 0.71 | — | — | 27 | 1.13 |
| Petroleum | 17 | 1.21 | 15 | 5.00 | 2 | 2.00 | — | — | 34 | 1.54 |
| Other manufacturing | 23 | 0.74 | 9 | 2.25 | 1 | 0.20 | 4 | 0.57 | 37 | 0.57 |
| Total | 178 | 1.72 | 79 | 2.26 | 168 | 4.54 | 100 | 4.00 | 530 | 2.51 |
| *By host country* | | | | | | | | | | |
| United States | — | — | 30 | 0.86 | 32 | 0.86 | 40 | 1.60 | 106 | 0.99 |
| United Kingdom | 39 | 0.38 | — | — | 15 | 0.41 | 6 | 0.24 | 61 | 0.35 |
| Other Europe | 71 | 0.68 | 28 | 0.80 | 90 | 2.43 | 27 | 1.08 | 216 | 1.02 |
| Japan | 31 | 0.30 | 5 | 0.14 | 14 | 0.38 | — | — | 50 | 0.27 |
| Other countries | 37 | 0.36 | 16 | 0.46 | 17 | 0.46 | 27 | 1.08 | 97 | 0.48 |
| Total | 178 | 1.72 | 79 | 2.26 | 168 | 4.54 | 100 | 4.00 | 530 | — |

*Source:* Pearce and Singh data base.

*Note:* Information was categorized as follows:

A. number of R&D laboratories from a particular parent country in industry or host country.

B. Number of R&D laboratories per parent MNE from a particular home country.

[a]Includes laboratories from one continental European country in another.

[b]Includes laboratories from other countries.

below-average number of units in Japan, the geographical dispersion of U.K. MNEs' R&D facilities appears quite typical, while petroleum is the industry with the most distinctive prevalence of overseas labs. The overall relatively below-average numbers of overseas laboratories per U.S. parent is also reflected in most industries, with electronics and pharmaceuticals again emerging as strongest.

The questionnaire sent to subsidiary laboratories in the Pearce and Singh survey provides evidence on the prevalence and characteristics of the three types of units outlined in the previous section. In table 2.3, SLs emerge as least prevalent, confirming that formally constituted overseas R&D units now play more ambitious roles than adaptive work, though some of the more

**TABLE 2.3. Prevalence of Particular Types of Overseas R&D Laboratories**

| | Average response[a] | | |
|---|---|---|---|
| | Support Laboratory (SL) | Locally Integrated Laboratory (LIL) | International Interdependent Laboratory (IIL) |
| *Industry* | | | |
| Food, drink, and tobacco | 1.91 | 2.73 | 1.64 |
| Petroleum | 2.20 | 2.00 | 2.40 |
| Metal manufacture and products | 1.73 | 2.30 | 1.46 |
| Industrial and agricultural chemicals | 2.06 | 2.06 | 2.11 |
| Pharmaceuticals and consumer chemicals | 1.46 | 1.62 | 2.69 |
| Motor vehicles (including components) | 1.25 | 2.50 | 1.75 |
| Industrial and farm equipment | 2.25 | 2.67 | 1.50 |
| Electronics and electrical appliances | 2.08 | 2.17 | 1.75 |
| Office equipment (including computers) | 1.25 | 1.63 | 2.38 |
| Other manufacturing | 1.75 | 2.20 | 1.80 |
| Total | 1.80 | 2.05 | 2.10 |
| *Host County* | | | |
| United States | 1.91 | 2.19 | 1.89 |
| United Kingdom | 1.62 | 1.88 | 2.32 |
| Other Europe | 1.96 | 1.96 | 2.22 |
| Japan | 1.50 | 2.00 | 3.00 |
| Other countries | 1.43 | 2.12 | 2.12 |
| Total | 1.80 | 2.05 | 2.10 |
| *Home country* | | | |
| United States | 1.67 | 2.02 | 2.12 |
| United Kingdom | 1.89 | 2.24 | 1.84 |
| Other Europe | 1.89 | 1.98 | 2.17 |
| Japan | 2.00 | 2.67 | 2.33 |
| Other countries | 1.61 | 2.00 | 2.00 |
| Total | 1.80 | 2.05 | 2.10 |

*Source:* Pearce and Singh 1992, table 5.1, 115.

*Note:* Responding subsidiary laboratories were asked to grade their facilities in terms of each type of unit as "predominantly this type of laboratory," "partially this type of laboratory," or "not this type of laboratory."

[a]Average response calculated by allocating values to the responses of 3 for "predominantly," 2 for "partially," and 1 for "not."

mundane SL functions may be done less formally in maintenance or engineering departments.[19] SLs were notably prominent in industrial and farm equipment and petroleum but rare in motor vehicles and office equipment. LILs are shown to be particularly prevalent in food, drink, and tobacco; industrial and farm equipment (which, taken with the similarly extensive presence of SLs, shows the locally responsive nature of this industry, where customization of

19. See Papanastassiou and Pearce 1994c, table 2, 802.

machinery plays a large role); and motor vehicles (compensating for the rarity of SLs in the industry). LILs were least prevalent in pharmaceuticals and office equipment, two industries where SLs were also relatively rare. They were of above-average relevance in the United States, which, taken with the similar position of SLs, shows the importance MNEs place on responding fully to the needs of that market. Subsidiary R&D units of Japanese and U.K. parent companies were most likely to do LIL work, which may show a particular lack of confidence in home-country innovations to meet foreign needs.

Of the facilities that said they were now LILs and also specified their age, 70 percent were set up before 1966, 23 percent between 1966 and 1980 and 7 percent since 1980. The comparable figures for SLs were 50 percent, 32 percent, and 18 percent, respectively. A plausible interpretation of these results is that most units were originally set up to do adaptive SL-type work and then evolved, through increasing capability and ambition (as argued in our earlier discussion), into the more development-oriented LIL work. Thus many of the SL units that were established earliest may now have evolved into LILs, while fewer of the newer ones have yet made the transition. This process of evolution is in line with the patterns discovered in the pioneering analysis of Ronstadt (1977, 1978).

IILs emerge in table 2.3 as being most distinctively prominent (compared to SLs and LILs) in two particularly high-technology industries—pharmaceuticals and office equipment (including computers)—and are also relatively important in petroleum. Subsidiary labs in Japan and the United Kingdom have a distinctive orientation to IIL work, while those in the United States are relatively least committed to it. Japanese MNEs are most likely to incorporate IIL work in their overseas laboratories, with UK enterprises least likely to do so. By comparison with SLs and LILs, units that said they were predominantly IILs were of relatively recent origin, with 44 percent set up before 1966, 34 percent between 1966 and 1980, and 22 percent since 1980. This data suggests that interest in the IIL type of role is of relatively recent origin and may perhaps also tend to be more often implemented by newly created labs rather than through evolution. Thus 80 percent of IILs had been set up as a "fresh installation for a specific purpose," compared with 62 percent of both SLs and LILs.

The facilities that described themselves as predominantly IILs emerged as particularly distinctive in the length of the projects they undertook.[20] Thus only 18 percent of IIL projects normally lasted fewer than 2 years, compared with 71 percent of SL projects and 67 percent of those in LILs. By contrast, 42 percent of IIL projects normally lasted more than 3 years, compared with 8

20. See Pearce and Singh 1992, table 5.6, 133.

percent of those in SLs and 14 percent in LILs. This data, in turn, reflects the distinctive prevalence of basic research in IILs, with 29 percent of such units reporting they did this work regularly, compared with 3 percent of LILs and no SLs. Thus of the overseas subsidiary laboratories that regularly did basic research and also specified their predominant lab type, 84 percent were IILs.

CHAPTER 3

# Internationalization of Technology in Japanese Multinationals

The survey by Granstrand and Sjolander (1992; see chapter 2) provides evidence of the perceived importance of an increasing internationalization of R&D among Japanese and other respondents (see table 3.1). Interestingly, the results show that despite a modest fall in the commitment to internationalization of *sales* by Japanese respondents, their commitment to internationalization of *production* rose notably. In light of our view that international production represents a potentially more proactive and responsive means of servicing international markets (compared with exports) and that it often needs increasing technological support within overseas operations to do so, this configuration of market orientation is compatible with the reported increased relevance of internationalized R&D in Japanese companies. A very similar evolution in market orientation (overseas production growing in perceived relevance more strongly than overseas sales) in Swedish companies again results in a notably increased commitment to internationalized R&D. By contrast, the relatively smaller increase in perceived importance of internationalized R&D in U.S. respondents appears alongside an increased internationalization of sales that embodies a much smaller rise in internationalization of production.

## Internationalization of R&D in Japanese MNEs

The aforementioned results, then, are broadly compatible with the view that internationalization of R&D in Japanese MNEs is an important element in the implementation of an approach to global competition that is increasingly based around international production. However, another finding reported by Granstrand and Sjolander (1992, 191) may add a further element to the explanation of emerging overseas R&D in these companies. They observe that almost all the Japanese companies "had diversified their technology bases significantly during the last decade, often in tandem with product diversification. However, even in cases of no product diversification, Japanese companies diversified their technology bases to a certain degree, due to increasing technological complexity in products and processes and advances in general technologies (microelectronics, new materials, etc.)." An effective broaden-

**TABLE 3.1. Perceived Importance of Corporate Development Strategies, 1982 and 1987**

| | Japan | | Sweden | | United States | | General | |
|---|---|---|---|---|---|---|---|---|
| | 1982 | 1987 | 1982 | 1987 | 1982 | 1987 | 1982 | 1987 |
| Internationalization of R&D | 1.7 | 2.7 | 1.4 | 2.9 | 1.5 | 2.4 | 1.6 | 2.6 |
| Internationalization of sales | 3.3 | 3.1 | 2.9 | 3.1 | 2.9 | 3.6 | 3.0 | 3.3 |
| Internationalization of production | 2.4 | 3.1 | 2.1 | 3.0 | 2.6 | 2.7 | 2.4 | 2.9 |
| Investments in R&D | 3.2 | 3.5 | 2.6 | 2.7 | 3.2 | 3.1 | 3.1 | 3.2 |
| Product diversification | 2.9 | 2.9 | 2.6 | 1.7 | 2.8 | 2.5 | 2.8 | 2.5 |
| Profitability | 3.0 | 3.5 | 2.9 | 3.4 | 3.5 | 4.0 | 3.2 | 3.7 |
| Growth | 3.4 | 3.6 | 2.6 | 2.7 | 3.2 | 3.1 | 3.1 | 3.2 |

*Source:* Granstrand and Sjolander 1992, table 9.4, 191.

*Note:* Scale ranges from 0 to 4, with 0 representing "of no importance" and 4 representing "of major importance."

ing of technological scope in this way may require scientific inputs from a range of countries and thus an expansion of overseas R&D of the IIL type.

The recent emergence of overseas R&D in Japanese MNEs can also be illustrated with evidence from the Pearce and Singh survey (1992; see chapter 2). The question in the survey of parent laboratories that asked them to list their R&D units by country also requested that they provide the dates of establishment of these facilities. Though this information was only available for 452 laboratories, the results, presented in table 3.2, provide some clear perspectives. Thus of the laboratories set up at home by the Japanese respondents to this question, 67.7 percent had been put in place before 1970, with 12.9 percent following during the 1970s and 19.4 percent in the 1980s. By contrast, none of their overseas labs were established before 1970, with 9.7 percent then created during the 1970s and the remainder in the subsequent decade. Therefore overseas facilities accounted for none of those set up by these firms before 1970, for 42.9 percent of those dating from the 1970s, and for 83.8 percent of those put in place in the 1980s. Further, while Japanese companies made no contribution to the 30.2 percent of all overseas labs established before 1970, they then contributed increasingly to the subsequent general expansion of overseas R&D. Thus 27.5 percent of all overseas laboratories were set up during the 1970s, with Japanese MNEs accounting for 5.8 percent of them, and 42.3 percent were created during the 1980s, with the Japanese share rising to 38.8 percent.

As a further reflection of the process of evolution of overseas R&D in MNEs, the Pearce and Singh survey asked parent laboratories to evaluate the

**TABLE 3.2. Date of Establishment of R&D Units**

| | Pre–1940 | | 1940–59 | | 1960–69 | | 1970–79 | | 1980–90 | |
|---|---|---|---|---|---|---|---|---|---|---|
| | Home | Overseas | Home | Overseas | Home | Overseas | Home | Overseas | Home | Overseas |
| Japan | 5 | 0 | 5 | 0 | 11 | 0 | 4 | 3 | 6 | 31 |
| United States | 29 | 2 | 39 | 11 | 27 | 14 | 13 | 20 | 25 | 19 |
| United Kingdom | 12 | 3 | 16 | 4 | 8 | 7 | 5 | 12 | 8 | 15 |
| Other Europe[a] | 10 | 4 | 11 | 3 | 7 | 8 | 4 | 16 | 5 | 14 |
| Other Countries[b] | 1 | 1 | 5 | 0 | 4 | 0 | 1 | 1 | 2 | 1 |
| Total | 57 | 10 | 76 | 18 | 57 | 29 | 27 | 52 | 46 | 80 |

*Source:* Pearce and Singh data base.

[a]Home country refers to individual country in Other Europe group, with other countries in this group included in Overseas.

[b]Home country refers to individual country in Other Country group, with other countries in this group included in Overseas.

**TABLE 3.3. Parent Laboratory Attitudes to Future R&D Strategy (in percentage)**

| | Parent Country | | | | |
|---|---|---|---|---|---|
| | Japan | United States | United Kingdom | Other Europe | Total |
| More use of the centralized facility | 7.4 | 18.1 | 33.3 | 21.1 | 20.2 |
| More emphasis on autonomous overseas laboratories | 11.1 | 8.4 | 19.4 | 15.8 | 12.8 |
| Increased emphasis on a globally integrated R&D network | 81.5 | 69.9 | 41.7 | 60.5 | 63.8 |
| No change | — | 3.6 | 5.6 | 2.6 | 3.2 |
| Total | 100.0 | 100.0 | 100.0 | 100.0 | 100.0 |

*Source:* Pearce and Singh data base.

relevance of four possible developments in approach. The results are summarized in table 3.3. The continued momentum in internationalization of R&D that is generally demonstrated by these replies is most clearly endorsed by the Japanese respondents. Thus these firms had the strongest tendency to suggest they would place "increased emphasis on a globally-integrated R&D network" (Pearce and Singh 1992, 70) and (as would be expected from their relatively recent implementation of geographically dispersed R&D facilities), notably, the least inclination to predict retrenchment to a centralized facility. This finding is compatible with the view that by coming relatively late to the setting up of overseas R&D units and by doing so at a time when most MNEs had already implemented a global approach to competition, these Japanese firms avoided the need for the traditional local-market-oriented role for these facilities. Instead Japanese MNEs were able to set up R&D laboratories abroad that supported the implementation of a globally integrated approach to innovation or took a position in an internationally coordinated network of basic research operations.[1]

## Origins of Patents of Japanese MNEs

The data on patenting in the United States by the world's leading enterprises, which we introduced in chapter 1, can also provide useful evidence on the position of overseas creative activity in Japanese MNEs. This data shows that in 1987–90 the overseas patents of Japanese companies (i.e., those attribut-

1. Another question in the Pearce and Singh survey asked parent companies that had no R&D facilities overseas if they had "recently considered initiating, or permitting the establishment of, a foreign R&D facility." A positive answer to this question was provided by 44 percent of relevant Japanese respondents, compared with 35 percent of those from the United States, 29 percent from continental Europe, and 18 percent from the United Kingdom.

able to R&D carried out in facilities outside of Japan) were 111 percent higher than they had been in 1969–72. Since this rise in overseas patenting of Japanese firms is considerably above the 30 percent increase for all the companies covered by the data, it results in an increase in their (still modest) share of all overseas patenting, from 1.5 percent to 2.5 percent (as reported in table 3.4). On the other hand, since the 111 percent rise in overseas patenting of the Japanese firms was well below the 498 percent increase in their overall patenting in the same period, the share of overseas patents in the total for these firms fell from 2.6 percent in 1969–72 to 0.9 percent in 1987–90 (see table 3.5).

The figures just reported suggest that much of the surge in the establishment of overseas laboratories by Japanese MNEs, that we note in other parts of our analysis, is too recent to have resulted in extensive patenting during the period covered. Nevertheless, the patent data already indicates a quite distinctive position for overseas R&D in Japanese MNEs. Thus the strong rise in overseas patenting that is discernible may suggest that the emerging foreign-based R&D operations in the Japanese enterprises are already playing a distinct *supporting* role within the even stronger firm-level momentum toward the creation and implementation of new technology. The expansion of overseas R&D in these enterprises therefore appears to have been playing a role that complements a parallel expansion of R&D operations in Japan and contributes to the implementation of the type of globally integrated approach to creativity and innovation we have outlined earlier.

The behavior in Japanese firms that we have just observed may well contrast somewhat with the developments in firms from three other coun-

**TABLE 3.4. Shares of Overseas Patents Granted to the World's Largest Firms in the United States by Home Country of Firm (in percentage)**

| | 1969–72 | 1973–77 | 1978–82 | 1983–86 | 1987–90 |
|---|---|---|---|---|---|
| Japan | 1.5 | 1.9 | 1.8 | 2.6 | 2.5 |
| United States | 34.2 | 34.1 | 35.1 | 35.7 | 32.0 |
| Europe (total) | 61.1 | 60.5 | 59.7 | 58.5 | 62.1 |
| European Community (total) | 48.4 | 44.6 | 43.7 | 46.1 | 50.8 |
| United Kingdom | 27.9 | 22.4 | 19.0 | 18.6 | 19.0 |
| Germany | 9.1 | 9.8 | 12.3 | 13.3 | 15.5 |
| France | 1.7 | 1.8 | 1.8 | 2.2 | 3.7 |
| Netherlands | 8.4 | 8.5 | 8.5 | 9.8 | 11.4 |
| Other Europe (total) | 12.7 | 15.9 | 16.0 | 12.3 | 11.3 |
| Sweden | 1.7 | 2.1 | 2.6 | 2.9 | 2.3 |
| Switzerland | 10.9 | 13.4 | 12.7 | 8.8 | 8.2 |
| Total | 100.0 | 100.0 | 100.0 | 100.0 | 100.0 |

*Source:* U.S. patent data base held at the University of Reading, with the support of the Patent and Trademark Office, U.S. Department of Commerce.

**TABLE 3.5. Share of Total U.S. Patents Granted to the World's Largest Firms Accounted for by their Overseas R&D, by Home Country of Firm (in percentage)**

| | 1969–72 | 1973–77 | 1978–82 | 1983–86 | 1987–90 |
|---|---|---|---|---|---|
| Japan | 2.6 | 1.9 | 1.2 | 1.3 | 0.9 |
| United States | 5.0 | 5.8 | 6.5 | 7.6 | 8.0 |
| Europe (total) | 28.8 | 25.9 | 24.6 | 27.1 | 29.5 |
| European Community (total) | 27.4 | 23.5 | 21.8 | 25.5 | 28.3 |
| United Kingdom | 45.4 | 43.9 | 41.1 | 47.9 | 49.2 |
| Germany | 12.8 | 11.0 | 12.1 | 14.5 | 17.1 |
| France | 8.2 | 7.7 | 7.2 | 9.2 | 15.9 |
| Netherlands | 50.4 | 47.4 | 47.7 | 54.0 | 53.9 |
| Other Europe (total) | 36.2 | 36.4 | 37.6 | 35.5 | 36.8 |
| Sweden | 18.0 | 19.5 | 25.2 | 28.8 | 28.3 |
| Switzerland | 44.5 | 43.6 | 43.8 | 41.6 | 43.0 |
| Total | 10.3 | 10.9 | 10.6 | 11.1 | 11.2 |

*Source:* U.S. patent data base held at the University of Reading, with the support of the Patent and Trademark Office, U.S. Department of Commerce.

tries—Germany, France, and the Netherlands—that also reported high rates of growth of overseas patenting during the period 1969–72 to 1987–90 (i.e., 121 percent, 192 percent, and 77 percent, respectively). Here, however, at least by contrast with the Japanese companies, these rates of growth of overseas patenting combine with lower growth in overall patenting (i.e., 66 percent, 51 percent, and 66 percent, respectively). These growth rates resulted in overseas patenting increasing its share of the overall patenting of German and French firms quite notably and also allowed it to retain its strong position in that of Dutch companies (table 3.5). Nevertheless it seems quite possible that in at least some of these cases, overseas R&D substitutes for work that would otherwise have been carried out at home. Thus in enterprises from these countries, where the overall growth of technological activity is much lower than for Japanese companies, it seems likely that the establishment of overseas R&D facilities may often represent *restructuring* of operations within a program whose overall magnitude is relatively unchanged. This interpretation can be contrasted with our view that in Japanese firms the overseas component of R&D is a distinctive increment in an R&D *growth* program.

For U.S. firms the level of overseas patenting in 1987–90 was 22 percent higher than it had been in 1969–72, while overall patenting was 24 percent lower, so that overseas patents increased their share from 5.0 percent to 8.0 percent (table 3.5). This data may suggest that among these companies, the momentum toward the internationalization of technological activity was able to assert itself to some degree, despite an overall bleak situation for creative operations. Finally, for U.K. companies, overseas patenting and total patenting were, respectively, 11 percent and 18 percent lower in 1987–90 than they had been in 1969–72, indicating that the overseas operations had been forced to share in a general retrenchment.

Table 3.6 shows that in 1987–90 overseas patenting was most strongly established among Japanese companies in the electrical equipment industry, which also, though by much smaller margins, accounted for most overseas patents of both U.S. and European firms. The office equipment industry claims quite strong positions in the overseas patenting of Japanese and U.S. firms, but not for those from Europe. The shares of the overseas patenting of Japanese and U.S. firms are also similar and quite strong in chemicals, but this time at levels well below that of European firms. Two industries that claim somewhat comparatively stronger positions among the overseas patenting of Japanese companies than U.S. or European are motor vehicles and instruments, while those where the Japanese position is distinctively weak are pharmaceuticals, petroleum and coal, and mechanical engineering.

## International Strategic Technology Alliances of Japanese MNEs

The data on STAs presented in table 1.12 may be further analyzed in order to discern the extent to which Japanese companies used interbloc (i.e., with European or U.S. companies) collaborations of this type as a means of ex-

**TABLE 3.6. Shares of Overseas Patents by Industry and Home Country of Firms, 1987–90 (in percentage)**

| | Home Country | | |
|---|---|---|---|
| Industry | Japan | United States | Europe[a] |
| Food, drink, and tobacco | 1.1 | 2.0 | 4.3 |
| Chemicals | 11.3 | 10.8 | 26.9 |
| Pharmaceuticals | 0.2 | 13.1 | 6.4 |
| Metals | 5.7 | 2.7 | 4.6 |
| Mechanical engineering | 0.7 | 6.7 | 4.7 |
| Electrical engineering | 35.1 | 17.8 | 28.2 |
| Office equipment | 17.9 | 16.8 | 1.1 |
| Motor vchicles | 13.6 | 7.6 | 5.4 |
| Rubber and plastic products | 1.1 | 1.8 | 0.2 |
| Nonmetallic minerals | 2.3 | 2.6 | 1.0 |
| Petroleum and coal | 0.5 | 7.8 | 14.3 |
| Professional and scientific instruments | 8.4 | 2.2 | 0.2 |
| Other | 2.3 | 7.9 | 2.7 |
| Total | 100.0 | 100.0 | 100.0 |

*Source:* U.S. patent data base held by the University of Reading, with the support of the Patent and Trademark Office, U.S. Department of Commerce.

[a]Includes patents deriving from R&D in European countries other than the home country of the MNE.

tending their technological scope. Overall Japanese firms are recorded in the MERIT-CATI data as having the strongest relative orientation toward interbloc (rather than intrabloc) STAs. Thus 85.2 percent of the STAs involving Japanese firms in 1980–84 were interbloc, as were 73.8 percent in 1985–89. The comparable figures for U.S. firms were 61.8 percent and 58.4 percent and for European firms, 63.4 percent and 57.5 percent. Looked at from another perspective (and bearing in mind the general below-average participation of Japanese firms in the recorded STAs), we find that Japanese companies were involved in 43.6 percent of all interbloc STAs in 1985–89, compared with 52.1 percent in 1980–84. While the United States kept its share constant (85.7 percent in 1985–89 compared with 86.0 percent in 1980–84), the share of European enterprises rose from 61.9 percent to 70.6 percent. This data

**TABLE 3.7. International Strategic Technology Alliances as a Proportion of Total Strategic Technology Alliances by Economic Bloc, 1980–89 (in percentage)**

| | Japan[a] | | Europe[b] | | United States[c] | |
|---|---|---|---|---|---|---|
| | 1980–84 | 1985–89 | 1980–84 | 1985–89 | 1980–84 | 1985–89 |
| Biotechnology | 82.0 | 69.2 | 57.3 | 60.3 | 45.2 | 49.9 |
| pharmaceuticals | 84.8 | 69.3 | 74.5 | 71.8 | 47.9 | 53.7 |
| agro-chemicals | 71.4 | 70.4 | 41.7 | 50.0 | 38.8 | 46.2 |
| other chemical | 100.0 | 60.0 | 53.3 | 33.3 | 46.7 | 29.2 |
| New materials | 81.6 | 47.0 | 61.8 | 63.0 | 75.0 | 63.0 |
| Information technology | 86.8 | 84.4 | 71.7 | 59.1 | 67.2 | 54.6 |
| computers | 92.6 | 96.4 | 78.0 | 81.1 | 66.1 | 53.5 |
| industrial automation | 84.7 | 85.7 | 69.7 | 63.8 | 64.6 | 50.9 |
| microelectronics | 93.4 | 91.3 | 83.3 | 60.6 | 72.7 | 55.6 |
| software | 80.0 | 76.0 | 40.6 | 51.7 | 55.9 | 42.1 |
| telecommunications | 66.7 | 72.1 | 68.5 | 56.9 | 61.6 | 70.2 |
| other | 87.0 | 59.4 | 77.1 | 61.7 | 76.0 | 71.1 |
| Other technologies | 85.2 | 84.7 | 51.3 | 54.9 | 70.5 | 68.7 |
| automotive | 88.9 | 88.7 | 47.1 | 69.0 | 90.0 | 78.8 |
| aviation/defense | 80.0 | 42.9 | 50.0 | 36.5 | 59.6 | 45.3 |
| chemical | 89.1 | 89.1 | 57.0 | 64.1 | 75.9 | 72.6 |
| food and beverages | 100.0 | 100.0 | 62.5 | 42.9 | 42.9 | 60.0 |
| heavy electric/energy | 69.2 | 57.1 | 40.6 | 46.4 | 66.7 | 81.3 |
| instruments | 89.5 | 91.7 | 60.0 | 65.8 | 76.5 | 72.7 |

*Source:* Hagedoorn and Schakenraad (1991, 69–73).

[a]International STAs of Japanese companies (i.e., with U.S. or European companies) divided by these International STAs plus intra-Japan STAs (i.e., those between Japanese companies).

[b]International STAs of European companies (i.e., with U.S. or Japanese companies) divided by these International STAs plus intra-European STAs (i.e., those between European companies).

[c]International STAs of U.S. companies (i.e., with Japanese or European companies) divided by these International STAs plus intra-United States STAs (i.e., those between U.S. companies).

reflects the notable rise of Europe-United States STAs, which were 71.9 percent higher in 1985–89 than in 1980–84, while Europe-Japan rose by 48.5 percent and United States-Japan by only 12.8 percent.

As table 3.7 shows, interbloc STAs retained a relatively strong position in the technological collaborations of Japanese information technology companies throughout the period 1980–89. However, the absolute numbers of STAs entered into by Japanese enterprises in this industry changed little between 1980–84 and 1985–89 (see table 1.12), contrasting with a notable rise in Europe-United States collaborations. As a result, Japanese firms participated in only 42.5 percent of new interbloc information technology STAs in 1985–89, compared with 54.6 percent in 1980–84. Interbloc STAs also sustained a prominent position in Japanese firms' technological collaborations in the other technologies sector (especially in automobiles, chemicals, food and beverages, and instruments). Though (by contrast with information technology) this position did encompass some growth in the absolute numbers of interbloc STAs of Japanese enterprises, it was again less than the growth of Europe-United States alliances in the industry (see table 1.12). Thus the share of interbloc STAs involving Japanese enterprises again fell, from 51.7 percent in 1980–84 to 45.2 percent in 1985–89.

Though new materials was the sector in which Japanese companies showed the greatest tendency to extend their use of STAs, table 1.12 reveals that this growth was manifested relatively more strongly through alliances between Japanese companies than with partners from Europe or the United States. Thus table 3.7 shows that the share of interbloc STAs of Japanese new materials companies fell substantially in 1985–89 compared with 1980–84. Nevertheless this was the only sector in which Japanese companies did increase their participation in all interbloc STAs, from 49.2 percent in 1980–84 to 54.8 percent in 1985–89. The modest increase in the number of biotechnology STAs of Japanese companies was rather more focused on intra-Japan alliances than those with U.S. or European firms (table 1.12), so that table 3.7 reveals a fall in the share of such interbloc STAs. The notable rise in Europe-United States biotechnology alliances (table 1.12) also results in a decline in Japanese participation in all interbloc STAs, from 46.3 percent in 1980–84 to 37.4 percent in 1985–89.

CHAPTER 4

# Japanese MNEs' R&D Operations in Europe

The first part of this chapter reviews the extent of Japanese MNEs' technological activity in Europe. Three pieces of evidence are produced that reflect on it: (1) data on numbers of R&D facilities, (2) indications of the output of these facilities, as reflected in patent data, and (3) information on strategic technology alliances involving European and Japanese enterprises. The second section of the chapter looks at Japanese R&D in Europe in more qualitative or behavioral terms, reviewing its role and motivation within the wider technological operations of the investing companies.

## Extent of Japanese Technological Activity in Europe

### R&D Facilities

The recent emergence of a strong commitment of Japanese MNEs to R&D operations in Europe has been clearly documented by JETRO data (see table 4.1), with a useful division of these laboratories into those that are associated with Japanese manufacturing operations in Europe and those that function independently of such linkages. In terms of the distinction outlined in the earlier section regarding roles of overseas R&D facilities, we would expect the associated facilities to mainly play the roles ascribed to SLs and LILs, while the independent facilities would be expected to be more likely to encompass the work of IILs.

As table 4.1 shows, the number of associated laboratories in Europe tripled between 1990 and 1992, with a further, though slower, rise of 10.7 percent in the next year. The United Kingdom reinforced its leading position with 40.0 percent of these associated labs in 1993, compared with 30.0 percent in 1990. By 1993 Germany was clearly established as the second-ranked host country to these laboratories, though its 15.2 percent share compared with 16.0 percent in 1990. Spain's perhaps surprising prominent 1990 share (22.0 percent) declined notably by 1993 (9.1 percent) and included a fall in absolute numbers during 1992. In a similar fashion, the share of France declined from 14.0 percent to 9.7 percent and also incorporated an absolute

**TABLE 4.1. Japanese R&D Facilities in Europe, by Country, 1990–93**

| | Associated Facilities[a] | | | | Independent Facilities[b] | | | |
|---|---|---|---|---|---|---|---|---|
| | 1990 | 1991 | 1992 | 1993 | 1990 | 1991 | 1992 | 1993 |
| United Kingdom | 15 | 33 | 54 | 66 | 9 | 14 | 18 | 20 |
| France | 7 | 11 | 21 | 16 | 3 | 6 | 8 | 10 |
| Germany | 8 | 17 | 25 | 25 | 6 | 11 | 13 | 19 |
| Netherlands | 3 | 4 | 5 | 10 | 0 | 0 | 1 | 3 |
| Belgium/Luxembourg | 3 | 6 | 7 | 13 | 1 | 2 | 3 | 5 |
| Ireland | 1 | 1 | 1 | 3 | 0 | 1 | 1 | 1 |
| Spain | 11 | 15 | 23 | 15 | 0 | 0 | 0 | 2 |
| Italy | 1 | 7 | 6 | 6 | 2 | 1 | 2 | 3 |
| Finland | 0 | 0 | 1 | 1 | 0 | 0 | 0 | 0 |
| Norway | 0 | 0 | 1 | 2 | 0 | 0 | 0 | 0 |
| Sweden | 1 | 1 | 3 | 4 | 0 | 0 | 0 | 1 |
| Denmark | 0 | 0 | 0 | 0 | 0 | 1 | 1 | 1 |
| Austria | 0 | 0 | 1 | 3 | 0 | 0 | 1 | 0 |
| Switzerland | 0 | 0 | 0 | 1 | 1 | 3 | 3 | 2 |
| Iceland | 0 | 1 | 1 | 0 | 0 | 0 | 0 | 0 |
| Total | 50 | 96 | 149 | 165 | 22 | 39 | 51 | 67 |

*Source:* JETRO 1993, table I.1, 3.
*Note:* Numbers represent number of R&D facilities in January of each year.
[a]R&D facilities associated with manufacturing facilities.
[b]R&D facilities with no direct association with manufacturing operations.

fall in 1992. Alongside the United Kingdom, the strongest rises are reported by the Benelux countries, which followed a modest rise in share from 6.0 percent in 1990 to 8.1 percent in 1992, with a more significant increase to 13.9 percent in 1993.

In table 4.2, a somewhat different perspective is provided by expressing the number of Japanese associated R&D facilities as a percentage of Japanese manufacturing subsidiaries. This data seeks to provide a broad indication of the proportion of those producing operations that *could* receive R&D support that actually did so. Despite the very notable rise in this ratio between 1990 and 1993, its absolute level may still seem quite low. One factor that may contribute to this perception is that where a Japanese MNE has a number of manufacturing subsidiaries in Europe, an associated R&D unit could support several, or all, of them. A particular case in which that could happen would be an SL backing up a whole network of RP subsidiaries. LIL-type associated labs may be considered the more likely to have a one-to-one relationship with an RPM operation.

The rise in the ratio, reported in table 4.2, from 9.7 percent in 1990 to 23.1 percent in 1993 can embody two significant tendencies. First, it seems likely that it reflects an increasing tendency for new Japanese manufacturing affiliates in Europe to incorporate such SL- or LIL-type facilities right from

**TABLE 4.2. Japanese Associated R&D Facilities as a Percentage of Japanese Manufacturing Facilities in Europe, 1990–93**

| | 1990 | 1991 | 1992 | 1993 |
|---|---|---|---|---|
| United Kingdom | 11.6 | 18.8 | 28.9 | 33.3 |
| France | 7.9 | 10.4 | 18.4 | 13.2 |
| Germany | 9.3 | 17.3 | 23.8 | 23.4 |
| Netherlands | 9.1 | 11.1 | 11.4 | 21.7 |
| Belgium/Luxembourg | 11.1 | 17.6 | 17.1 | 31.0 |
| Spain | 20.4 | 23.4 | 36.5 | 23.8 |
| Italy | 3.6 | 18.4 | 13.3 | 13.3 |
| Sweden | 16.7 | 16.7 | 33.3 | 44.4 |
| Austria | 0 | 0 | 6.7 | 17.6 |
| Total | 9.7 | 15.2 | 21.7 | 23.1 |

*Source:* JETRO 1993, table I.1, 3.

their initiation. Since the number of manufacturing affiliates reported by JETRO rose from 515 in 1990 to 713 in 1993, there is clearly adequate scope for this factor to influence the overall ratio. Second, the rise in the ratio is also likely to include a growing tendency for manufacturing subsidiaries that had been set up earlier without the support of associated R&D laboratories to now feel the need to incorporate such facilities. Overall the rise in the ratio can be seen as a clear response by the Japanese MNEs to increased competition in the European market, which requires them to provide their manufacturing affiliates there with more of the technological support that will allow them to achieve a distinctive reaction to the needs and potentials of this challenging new environment.

Table 4.1 indicates that almost three-quarters of Japanese companies' independent laboratories in Europe in 1993 were in the United Kingdom (29.9 percent), Germany (28.4 percent), or France (14.9 percent). The overall growth of these independent laboratories has quite closely matched that of associated labs, so that table 4.3 reveals that in 1993 almost 30 percent of all Japanese units in Europe were still of this type. If we are justified in the view that these labs predominantly do basic precompetitive research mainly relating to the first phase of a globalized approach to innovation, while the associated labs mainly implement phase two, then it is clear that Japanese MNEs see the expansion of technological activity in Europe as encompassing a full-range contribution to their overall needs and programs.

### Patents from European Research

The data on patenting in the United States, already discussed in earlier sections, now helps us provide some perspectives on the emergence of creative

**TABLE 4.3. Independent Japanese R&D Facilities as a Percentage of Total Japanese R&D Facilities in Europe, 1990–93**

| | 1990 | 1991 | 1992 | 1993 |
|---|---|---|---|---|
| United Kingdom | 37.5 | 29.8 | 25.0 | 23.3 |
| France | 30.0 | 35.3 | 27.6 | 38.5 |
| Germany | 42.9 | 39.3 | 34.2 | 43.2 |
| Belgium/Luxembourg | 25.0 | 25.0 | 30.0 | 27.8 |
| Italy | 66.7 | 12.5 | 25.0 | 33.3 |
| Total | 30.6 | 28.9 | 25.5 | 28.9 |

*Source:* JETRO 1993, table I.1, 3.

activity in Japanese firms' operations in Europe.[1] Thus table 4.4 shows that in 1969–72, 26.3 percent of the patents taken out in the United States by Japanese companies as a result of their research outside of Japan were attributed to their European operations. Of these European patents Germany accounted for 37.1 percent and the United Kingdom for 28.6 percent. The subsequent period was generally one of greatly increased overseas patenting by Japanese MNEs, with these 77.4 percent higher in 1973–77 than they had been in 1969–72.[2] This rise, however, was strongly focused on the Japanese companies' U.S. operations, whose patenting increased by 110.0 percent, compared with only 5.1 percent for those in Europe, so that Europe's share of these companies' total overseas patents fell to 15.6 percent. Germany consolidated its position in the patenting from Japanese firms in Europe in the period 1973–77, with 52.2 percent of the total, compared with only 15.2 percent for the United Kingdom. The period 1978–82 was generally one of relatively low patenting,[3] probably reflecting the outcome of reductions in R&D expenditure implemented during the difficult economic conditions of the mid-1970s. Thus Japanese patents deriving from research in the United States rose by only 4.3 percent in this period, compared with the previous five years, while those originating from research in Europe fell by 39.1 percent. Within Europe, Germany, though sharing to some extent in the decline, clearly retained its leading position, accounting for 64.3 percent of Japanese patents deriving from research in the region.

1. Tables 4.4 and 4.6 exclude data on the nonmetallic minerals industry because exceptionally high patenting attributed to European operations of Japanese firms in this industry in 1969–77 greatly distorts the picture of more pervasive general trends.

2. The comparisons between 1973–77 and 1969–72 are based on the annual average patents in the two periods.

3. Thus the total patenting of Japanese firms in 1978–82 was only 27.2 percent higher than in 1973–77, a small rise in the context of an overall growth of 497.6 percent in the full period 1969–72 to 1987–90. For U.S. firms, patenting was actually 19.8 percent lower in 1978–82 than in 1973–77.

**TABLE 4.4. The Share of the U.S. Patents Granted to Leading Japanese Companies as a Result of their Overseas Research, by Host Country or Region (in percentage)**

| | 1969–72 | 1973–77 | 1978–82 | 1983–86 | 1987–90 |
|---|---|---|---|---|---|
| United States | 66.2 | 78.3 | 87.0 | 88.3 | 77.3 |
| Europe (total) | 26.3 | 15.6 | 10.1 | 11.5 | 19.7 |
| European Community (total) | 22.6 | 14.9 | 7.6 | 10.9 | 18.5 |
| United Kingdom | 7.5 | 2.4 | 0.4 | 7.3 | 15.3 |
| Germany | 9.8 | 8.1 | 6.5 | 2.2 | 1.6 |
| France | 0.8 | 1.7 | 0.4 | 0.3 | 0.2 |
| Other European Community | 4.5 | 2.7 | 0.4 | 1.1 | 1.4 |
| Non-EC Europe | 3.8 | 0.7 | 2.5 | 0.6 | 1.2 |
| Total | 100.0 | 100.0 | 100.0 | 100.0 | 100.0 |

*Source:* U.S. patent data base held at the University of Reading, with the support of the Patent and Trademark Office, U.S. Department of Commerce.

Japanese overseas patenting recovered generally in 1983–86, being 65.9 percent higher in this period than in 1978–82.[4] This growth was rather stronger in Europe, where patents resulting from Japanese firms' operations were 83.0 percent higher in 1983–86 than they had been in 1978–82, than in the United States, where the comparable increase was 63.9 percent. Though the growth in Europe-originated patenting of Japanese MNEs in 1983–86 restored the level to one that was only 11.4 percent above that of 1973–77, it does seem that this period of recovery in fact involved a significant reorientation of Japanese creative activity in Europe. It may therefore perhaps be that this period can be seen as the point of origination of the current phase of Japanese technological activity in Europe. With regard to the national distribution of Japanese MNEs' European patents, this reorientation provided a much stronger position for United Kingdom-based operations. Thus in 1983–86 Japanese MNEs' United Kingdom-derived patents were 364.3 percent higher than they had been in 1973–77, while those stemming from research in Germany were 58.4 percent lower. Reflecting that, the United Kingdom accounted for 63.4 percent of the total European patents of these companies in the period.

The trends just observed persisted strongly in 1987–90. Thus while United States-derived patents of Japanese MNEs were only 5.7 percent higher in 1987–90 than in 1983–86, those originating from creative activity in Europe increased by 107.3 percent. Once again, research in the United Kingdom took a central position in the Japanese MNEs' European operations, with United Kingdom-derived patenting increasing by 153.8 percent between 1983–86 and 1987–90 and accounting for 77.6 percent of these firms' Eu-

4. The comparisons between 1983–86 and 1978–82 and 1973–77 are based on the annual average patents in the three periods.

ropean-derived patents in the latter period. Overall Japanese MNEs' European operations accounted for 19.7 percent of all their overseas patents in 1987–90, compared with 11.5 percent in 1983–86 and 10.1 percent in 1978–82.

Aspects of the industrial distribution of Japanese MNEs' research in Europe, as reflected in patent data, are reported in tables 4.5 and 4.6. Electrical engineering has been the key industry with respect to Japanese MNEs' patenting from research carried out in Europe. Over the whole period 1969–90, this industry accounted for 45.5 percent of all Japanese companies' Europe-derived patents, while these European patents were 25.2 percent of all those resulting from overseas R&D in Japanese electrical engineering enterprises. However, as the tables show, both these ratios are notably at their highest in the period 1983–90. Thus if we argue, as indicated, that the origins of the current wave of Japanese MNEs' R&D in Europe may be located in the 1983–86 period, then it is clear that a commitment to such R&D by electrical engineering companies is a key characteristic of this phase. In fact, this characteristic then clearly parallels another central characteristic of this wave of R&D, namely, its focus on operations in the United Kingdom. Thus the United Kingdom accounted for only 20.8 percent of Japanese electrical engineering MNEs' Europe-originated patents in 1969–82, but for 90.4 percent of those in 1983–90, with Germany accounting for 58.3 percent in the former period but only 7.2 percent in the latter.

The other industry in which Japanese MNEs are revealed, through patent data, as having had a distinctive and sustained commitment to Europe-based R&D is professional and scientific instruments. Over the period 1969–90 the instruments industry accounted for 24.2 percent of all patents resulting from Japanese companies' R&D in Europe, compared with only 5.9 percent of those deriving from their R&D in the United States. Reflecting this, table 4.5 shows that European patents accounted for more than 40 percent of all overseas patents of Japanese MNEs in this industry in each period covered

**TABLE 4.5. European Share of the U.S. Patents Granted to Leading Japanese Companies as a Result of their Overseas Research, by Industry (in percentage)**

| | 1969–72 | 1973–77 | 1978–82 | 1983–86 | 1987–90 |
|---|---|---|---|---|---|
| Chemicals | 8.2 | 3.8 | 2.2 | 1.3 | 6.0 |
| Electrical engineering | 28.1 | 8.5 | 15.0 | 29.0 | 33.5 |
| Office equipment | 64.3 | 3.3 | 2.9 | 0 | 0 |
| Motor vehicles | 36.4 | 41.7 | 7.1 | 3.0 | 10.0 |
| Professional and scientific instruments | 54.5 | 52.6 | 40.7 | 22.7 | 40.5 |

*Source:* U.S. data base held at the University of Reading, with the support of the Patent and Trademark Office, U.S. Department of Commerce.

**TABLE 4.6. Shares of the U.S. Patents Granted to Leading Japanese Companies as a Result of their Research in Europe and the United States, by Industry, 1969–90**

| | Europe | | United States | |
|---|---|---|---|---|
| | 1969–82 | 1983–90 | 1969–82 | 1983–90 |
| Chemicals | 8.3 | 3.2 | 36.6 | 19.2 |
| Electrical engineering | 22.0 | 65.9 | 24.1 | 27.1 |
| Office equipment | 10.1 | 0.0 | 11.6 | 20.5 |
| Motor vehicles | 10.1 | 6.3 | 7.0 | 17.8 |
| Professional and scientific instruments | 33.9 | 15.9 | 5.7 | 6.0 |
| Other | 15.6 | 8.7 | 15.0 | 9.4 |
| Total | 100.0 | 100.0 | 100.0 | 100.0 |

*Source:* U.S. data base held at the University of Reading, with the support of the Patent and Trademark Office, U.S. Department of Commerce.

except 1983–86. In fact, the slump in share in 1983–86 may reflect a transitional period between two phases of Japanese R&D in Europe in the instruments industry, with the United Kingdom again playing an enhanced role in the subsequent regeneration of activity. Thus Germany accounted for 62.2 percent and the United Kingdom for 13.5 percent of Japanese MNEs' total Europe-derived patents in instruments in 1969–82, while in 1983–90 the position reversed, with the United Kingdom taking 60.0 percent and Germany 20.0 percent.

Though Japanese office equipment MNEs derived a significant level of patents from their European activities in 1969–72, this situation was then superseded by an apparent decision to focus overseas research in this industry in the United States. Similarly, Europe occupied a relatively strong position in the overseas R&D of Japanese motor vehicle firms from 1969–77, with the United States once again then taking a notably stronger position. In the chemicals industry Germany persistently attracted the majority of the relatively limited amounts of Japanese R&D in Europe, accounting for 53.8 percent of all patents deriving from these companies' European activity in the full 1969–90 period.

## Strategic Technology Alliances

The MERIT-CATI data on strategic technology alliances (discussed earlier) allows us to indicate the position of collaborations with European enterprises among the total STAs of Japanese companies. As table 4.7 shows, the 48.5 percent growth in Europe-Japan STAs between 1980–84 and 1985–89 promoted pharmaceuticals, new materials, microelectronics, telecommunica-

**TABLE 4.7. Europe-Japan Strategic Technology Alliances, by Industry, 1980–89**

| | Total Number of Europe-Japan Strategic Technology Alliances | | R&D-driven STAs as a Proportion of all Europe-Japan STAs | |
|---|---|---|---|---|
| | 1980–84 | 1985–89 | 1980–84 | 1985–89 |
| Biotechnology | 5 | 20 | 60.0 | 80.0 |
| pharmaceuticals | 2 | 12 | 50.0 | 75.0 |
| agro-chemicals | 2 | 6 | 50.0 | 83.3 |
| other chemicals | 1 | 2 | 100.0 | 100.0 |
| New materials | 15 | 23 | 46.7 | 47.8 |
| Information technology | 57 | 57 | 50.9 | 45.6 |
| computers | 8 | 2 | 0 | 0 |
| industrial automation | 27 | 10 | 70.4 | 40.0 |
| microelectronics | 7 | 14 | 42.9 | 35.7 |
| software | 1 | 6 | 0 | 100.0 |
| telecommunications | 0 | 13 | — | 46.2 |
| other | 14 | 12 | 50.0 | 41.7 |
| Other technologies | 24 | 50 | 20.8 | 38.0 |
| automotive | 6 | 16 | 66.7 | 50.0 |
| aviation/defense | 1 | 0 | 0 | — |
| chemicals | 14 | 21 | 7.1 | 19.0 |
| food and beverages | 2 | 2 | 0 | 0 |
| heavy electric/energy | 0 | 4 | — | 25.0 |
| instruments | 0 | 6 | — | 100.0 |
| Total | 101 | 150 | 43.6 | 48.0 |

*Source:* Hagedoorn and Schakenraad 1991, 69–73.

tions, and automobiles into leading positions among such collaborations, while industrial automation notably lost ground from its preeminent position in the earlier period. Europe-Japan STAs improved their position among all interbloc STAs of Japanese companies, accounting for 32.7 percent in 1985–89, compared with 26.9 percent in 1980–84 (table 4.8). However, their share of all interbloc STAs of European firms fell somewhat, to 20.2 percent from 22.6 percent.

Among Europe-Japan STAs, those in new materials are shown in table 4.8 to have held the strongest sustained relative position, accounting for an above-average share of both Japanese and European companies' total interbloc collaborations in the industry in both periods. Europe-Japan STAs improved their relative position most notably in the rather disparate other technologies sector, accounting for above-average shares of all interbloc collaborations of both European and Japanese firms in the sector in 1985–89, compared with below-average shares in 1980–84. Europe-Japan STAs also improved their relative position among the interbloc STAs of both European and Japanese

**TABLE 4.8. Europe-Japan Strategic Technology Alliances as a Proportion of Total International Strategic Technology Alliances of Europe and Japan, by Industry, 1980–89**

| | Europe-Japan STAs as a Proportion of All International STAs of European Firms | | Europe-Japan STAs as a Proportion of All International STAs of Japanese Firms | |
|---|---|---|---|---|
| | 1980–84 | 1985–89 | 1980–84 | 1985–89 |
| Biotechnology | 7.9 | 13.9 | 10.0 | 27.0 |
| pharmaceuticals | 5.7 | 12.8 | 5.1 | 23.1 |
| agro-chemicals | 10.0 | 14.3 | 20.0 | 31.6 |
| other chemicals | 12.5 | 25.0 | 100.0 | 66.7 |
| New materials | 31.9 | 30.7 | 48.4 | 36.5 |
| Information technology | 26.5 | 18.2 | 30.0 | 30.2 |
| computers | 25.0 | 6.7 | 32.0 | 7.4 |
| industrial automation | 58.7 | 22.7 | 54.0 | 33.3 |
| microelectronics | 11.7 | 35.0 | 9.9 | 22.2 |
| software | 7.7 | 8.0 | 12.5 | 31.6 |
| telecommunications | 0 | 14.9 | 0 | 41.9 |
| other | 51.9 | 32.4 | 70.0 | 63.2 |
| Other technologies | 19.8 | 23.7 | 23.1 | 37.6 |
| automotive | 37.5 | 40.0 | 37.5 | 29.1 |
| aviation/defense | 4.0 | 0 | 12.5 | 0 |
| chemicals | 31.1 | 28.0 | 28.6 | 42.9 |
| food and beverages | 40.0 | 33.3 | 100.0 | 50.0 |
| heavy electric/energy | 0 | 15.4 | 0 | 50.0 |
| instruments | 0 | 24.0 | 0 | 54.5 |
| Total | 22.6 | 20.2 | 26.9 | 32.7 |

*Source:* Hagedoorn and Schakenraad 1991, 69–73.

enterprises in biotechnology but still remained of below-average significance. Finally, these Europe-Japan STAs retained their position among the total interbloc collaborations of Japanese firms in information technology but lost ground notably among those of European companies.

Though R&D-driven collaborations increased their share of Europe-Japan STAs from 43.6 percent in 1980–84 to 48.0 percent in 1985–89, they still were left slightly less prevalent than in Japan-United States STAs (47.8 percent and 52.1 percent in the two periods) and distinctively less prevalent than in Europe-United States STAs (51.3 percent and 65.3 percent). Table 4.7 indicates that R&D-driven Europe-Japan STAs are most common in the biotechnology sector and rarest in other technologies (notably chemicals).

An analysis by Mowery and Teece (1992, table 6) of MITI data on 135 international research joint ventures[5] involving Japanese firms initiated between 1982 and 1987 showed that 93 of these were with U.S. partners, and 34 were with European companies (United Kingdom 13, Germany 9, France 7, Italy 5).[6] The data also showed that Japanese research joint ventures with European partners were relatively rare in advanced technology sectors,[7] where U.S. firms accounted for 85.1 percent of the total and those from the European countries for only 10.8 percent. In conventional technology sectors, however, European countries provided the Japanese companies' partners in 42.6 percent of the research joint ventures and the United States in 49.2 percent of them. These results need not necessarily be taken to imply that Japanese companies did not place a significant value on the general capability of the overall European science base in the more advanced areas of technology.

Since other results[8] showed that the majority of these joint ventures focused on applied research and development, the result for Europe may perhaps sometimes reflect a view by Japanese companies that European firms were not themselves making an adequate job of exploiting the underlying potentials of European science in these areas. In this case it might then be that the Japanese MNEs were opting instead to access European technology in these sectors more directly, through the establishment of their own facilities for both development and applied research, rather than relying on picking it up as, at least partly, embodied in the firm-level research capabilities of European enterprises. There are then two corollaries of this argument, as implied by the data on Japanese research joint ventures. The first is that the Japanese companies seem to be more inclined to trust their U.S. joint venture partners to have accessed U.S. technological capabilities more effectively in the advanced technology sectors, so that they, in turn, can benefit from this broad expertise through the firm-level collaborations. The second implied corollary is that Japanese enterprises were demonstrating somewhat greater faith in European research joint venture partners to have more effectively tapped into European technology's potentials in the more conventional areas of science.

5. Mowery and Teece (1992, 124) note that the MITI "report appears to have employed a fairly narrow, legalistic definition of a joint venture."

6. The remainder were with Canadian companies (4) or other unspecified sources.

7. Most of the research joint ventures in advanced technology sectors were in communications and computers, integrated circuits, factory and office automation, and biotechnology.

8. Analysis of a slightly larger number of cases (Mowery and Teece 1992, table 7) showed that 46 percent of Japanese firms' research joint ventures in advanced technology sectors were "product oriented," 28 percent were "market specified" (i.e., related to "incremental product modifications for new markets"), and 22 percent involved "applied research." In conventional technology sectors, the shares of these types of work were 35 percent, 43 percent, and 11 percent, respectively.

## Motivations and Roles of Japanese Firms' European R&D

In this section the results of recent surveys of Japanese companies' European subsidiaries, carried out annually by JETRO, are discussed in order to provide evidence on both the broader R&D strategies of these MNEs and the particular motives and roles of R&D within the European operations.

### R&D Strategy in Japanese MNEs

The views of Japanese MNEs' European subsidiaries on the likely future of R&D strategies of their groups are reported in table 4.9. Though these replies may to some extent reflect the wishes of these Europe-based subsidiary-level respondents, rather than actual declared central-managerial intentions, they also appear to clearly indicate further support for the view of a strong momentum toward decentralization of R&D as part of a globalized approach to the creation and use of technology. Thus only 21.2 percent of these European subsidiaries felt that the most plausible future R&D system of their group was to centralize all strategic technological activities such as basic research and the development of new products.

Of the two approaches that embody decentralized R&D activity, 55.2 percent of the European subsidiaries endorsed the purely demand-related influence that design facilities would be used to meet local needs. This approach provides subsidiary-level R&D with a distinct, if limited, role within the more contemporary type of globally competitive market-seeking FDI, with its aims essentially constrained to the local application of, perhaps recently created, established technology. However, 23.7 percent of the respondents anticipated a more broadly based role for decentralized R&D in these Japanese MNEs, indicating that "international bases will be installed, including basic research and design centers, in order to globalize the company's activities." Once again this approach provides a demand-side product-development role for the R&D activity, though the prospect of this element being incorporated within such international bases may imply its deeper integration with basic and applied research in a manner that could provide subsidiaries with the more distinctive technological independence of RPMs or WPMs. In addition, however, these international bases are envisaged to include scope for basic research as part of global activities, perhaps in the manner proposed for the first phase of a global-innovation strategy. Therefore supply-side factors also enter in this international R&D bases system, with the location and scientific specialization of such laboratories reflecting the capacities of host countries' science bases, in a manner that reflects the more contemporary knowledge-oriented type of resource-seeking FDI.

**TABLE 4.9. Future R&D Systems of Japanese MNEs as seen by their European Subsidiaries**

| | Number of Replies | Future System (%) | | | |
|---|---|---|---|---|---|
| | | A | B | C | Total |
| *By Host Country* | | | | | |
| United Kingdom | 126 | 20.6 | 49.2 | 30.2 | 100.0 |
| France | 42 | 28.6 | 54.8 | 16.7 | 100.0 |
| Germany | 57 | 17.5 | 61.4 | 21.1 | 100.0 |
| South Europe[a] | 51 | 17.6 | 70.6 | 11.8 | 100.0 |
| North Europe[b] | 25 | 32.0 | 32.0 | 36.0 | 100.0 |
| Benelux | 53 | 18.9 | 62.3 | 18.9 | 100.0 |
| Others[c] | 5 | 20.0 | 20.0 | 60.0 | 100.0 |
| Total | 359 | 21.2 | 55.2 | 23.7 | 100.0 |
| *By Sector* | | | | | |
| Assembling and processing (total) | 125 | 23.2 | 56.8 | 20.0 | 100.0 |
| General machinery | 27 | 29.6 | 63.0 | 7.4 | 100.0 |
| Electronic and electrical equipment | 70 | 25.7 | 54.3 | 20.0 | 100.0 |
| Transport equipment | 12 | 8.3 | 58.3 | 33.3 | 100.0 |
| Precision instruments | 16 | 12.5 | 56.3 | 31.3 | 100.0 |
| Parts manufacturing (total) | 72 | 23.6 | 56.9 | 19.4 | 100.0 |
| Electronic components | 44 | 25.0 | 61.4 | 13.6 | 100.0 |
| Transport equipment parts | 28 | 21.4 | 50.0 | 28.6 | 100.0 |
| Materials purchasing | 72 | 26.4 | 58.3 | 15.3 | 100.0 |
| Chemical products | 54 | 22.2 | 55.6 | 22.2 | 100.0 |
| Others | 25 | 4.0 | 72.0 | 24.0 | 100.0 |
| Design centers and R&D bases | 35 | 2.9 | 34.3 | 62.9 | 100.0 |
| Total[d] | 383 | 20.6 | 55.9 | 23.5 | 100.0 |

*Source:* JETRO 1993, table V.1, 40.
*Note:* Information on R&D systems was categorized as follows:

A. The parent company will centralize all the activities, such as basic research and the development of new products.
B. Some of the design facilities will be left to meet the local needs.
C. International bases will be installed, including basic research and design centers, in order to globalize the company's activities.

[a]Italy, Spain, Portugal, and Greece.
[b]Finland, Norway, Sweden, Denmark, and Ireland.
[c]Austria, Switzerland, and Iceland.
[d]Some of the 359 respondents were diversified and are reported in two sectors.

The respondents that did anticipate a persisting centralization of technological activity in these Japanese MNEs were asked to evaluate three possible reasons for this approach. Perhaps the result of clearest interest to our argument is that only 14.7 percent of replies (see table 4.10) endorsed the reason "Japanese conceptions and designing abilities are sufficient to carry on the R&D activities." Thus even where other factors appear to preclude overseas R&D, there appears to be extensive acknowledgment that benefits could

**TABLE 4.10. Reasons for Centralizing the R&D System in the Parent Company as Seen by Japanese Operations in Europe**

| | Number of Responding Affiliates | Reason for Centralization (in percentage of responding affiliates that endorsed a particular reason) | | | |
|---|---|---|---|---|---|
| | | A | B | C | D |
| *By Host Country* | | | | | |
| United Kingdom, France, and Germany | 47 | 36.2 | 14.9 | 72.3 | 6.4 |
| South Europe[a] | 9 | 33.3 | 11.1 | 77.8 | 22.2 |
| North Europe[b] | 8 | 62.5 | — | 50.0 | 12.5 |
| Benelux | 10 | 30.0 | 20.0 | 50.0 | 10.0 |
| Others | 1 | — | 100.0 | — | — |
| Total | 75 | 37.3 | 14.7 | 66.7 | 9.3 |
| *By Sector* | | | | | |
| Assembling and processing | 29 | 31.0 | 17.2 | 72.4 | 6.9 |
| Parts manufacturing | 17 | 41.2 | 11.8 | 64.7 | 11.8 |
| Material purchasing | 19 | 57.9 | 15.8 | 47.4 | 5.3 |
| Chemical products | 11 | 27.3 | 9.1 | 90.9 | 9.1 |
| Others | 1 | 100.0 | — | — | — |
| Design centers and R&D bases | 1 | — | — | — | 100.0 |
| Total[c] | 78 | 39.7 | 14.1 | 65.4 | 9.0 |

*Source:* JETRO 1993, table V.3.

*Note:* Reasons for centralization were categorized as follows:

A. Priority has been given to the development of the company's own technologies in order for the company to grow.

B. Japanese conceptions and designing abilities are sufficient to carry on the R&D activities.

C. There is no capability of decentralizing financial, human, and other resources.

D. Others.

[a]Italy, Spain, Portugal, and Greece.

[b]Finland, Norway, Sweden, Denmark, and Ireland.

[c]Three of the 75 respondents were diversified and are reported in two sectors.

accrue from the pursuit of quality specialized inputs and product development work in other countries. Contrasting with that, a perhaps more ideological commitment to the virtues of centralized technology creation and innovation may be enshrined in the view that "priority has been given to the development of the company's own technologies in order for the company to grow." This reason was endorsed by 37.3 percent (30) of the relevant respondents in table 4.10, but, therefore, by only 8 percent of the wider sample covered by table 4.9. Finally 66.7 percent of the respondents reported in table 4.10 indicated that centralization resulted because "there is no capability of decentralizing financial, human, and other resources," which suggests that the main limits on the adoption of more globalized approaches in these companies were perceived as being organizational difficulties or resource constraints.

## Motives for European R&D of Japanese MNEs

Respondents to the JETRO survey that did perceive a role for localized R&D facilities in their European operations were asked to evaluate the relevance of a number of possible motives for such activities. As table 4.11 shows, the view that "it is necessary to supply products meeting local needs" was seen to be of extremely pervasive relevance. Once again a creative subsidiary role within a contemporary approach to market-seeking FDI is strongly endorsed. In addition, however, an explicit motivation for the increasingly globalized approach to innovation that we envisage is also quite prevalent, in the suggestion that "it is necessary to reduce the lead time from R&D to commercialization." Thus many of the respondents do seem to perceive responsiveness to local needs as part of the innovation process itself, rather than as merely the more conventional and limited adaptation of established products.

Though table 4.11 indicates strong support for both traditional (adaptation) and contemporary (development) demand-related motives for European R&D in Japanese MNEs, it also provides distinct evidence of the emergence of supply-side influences. Thus the second most popular motive for decentralized R&D is that "it is necessary to know local technological trends and adapt to the greater technological competition as soon as possible." This response clearly indicates that these Japanese MNEs now see a worldwide perspective as a very decisive element in their approach to technological competition and that the European R&D facilities embody a "listening-post" function among their roles. Enhancing this perception of the European laboratories as supporting the emergence of a knowledge-focused type of resource-seeking FDI in Japanese companies is the relatively strong endorsement of "it is necessary to employ foreign researchers and use their ideas and conceptions in order to expand the scope of R&D" among the supply-side motives covered in table 4.11. Thus the increasing differentiation of national specializations in scientific capacities is verified as an important factor generating the growing globalization of firm-level research operations. The positive qualitative aspect of this decentralization is further demonstrated by the fact that few Japanese firms believed that "there is a shortage of researchers and developers in Japan" constituted a significant negative quantitative motive.

## Roles of European Laboratories of Japanese MNEs

A JETRO survey investigation of the types of work performed in Japanese MNEs' European-based R&D units is reported in table 4.12. Within the context of the widening range of motivations for the implementation of European R&D as part of a globalized pursuit of competitiveness, discussed in the previous section, these results indicate that types of work that focus on

**TABLE 4.11. Reasons for Localizing R&D Facilities, as Seen by Japanese Operations in Europe**

| | Number of Responding Affiliates | Reason for Localization (in percentage of responding affiliates that endorsed a particular reason) | | | | | | | | |
|---|---|---|---|---|---|---|---|---|---|---|
| | | A | B | C | D | E | F | G | H | I |
| *By Host Country* | | | | | | | | | | |
| United Kingdom, France, and Germany | 170 | 84.7 | 36.5 | 16.5 | 61.2 | 41.8 | 10.6 | 17.6 | 2.9 | 5.9 |
| South Europe[a] | 42 | 95.2 | 11.9 | 7.1 | 61.9 | 35.7 | 14.3 | 4.8 | 2.4 | 7.1 |
| North Europe[b] | 17 | 47.1 | 41.2 | 11.8 | 29.4 | 47.1 | 17.6 | 11.8 | — | 5.9 |
| Benelux | 43 | 90.7 | 39.5 | 14.0 | 65.1 | 34.9 | 4.7 | 11.6 | — | — |
| Others[c] | 4 | 50.0 | — | 25.0 | 50.0 | 75.0 | — | — | — | 25.0 |
| Total | 276 | 84.4 | 33.0 | 14.5 | 59.8 | 40.6 | 10.5 | 14.1 | 2.2 | 5.4 |
| *By Sector* | | | | | | | | | | |
| Assembling and processing | 93 | 90.3 | 30.1 | 15.1 | 73.1 | 37.6 | 11.8 | 20.4 | 4.3 | 5.4 |
| Parts manufacturing | 55 | 87.3 | 27.3 | 5.5 | 50.9 | 50.9 | 18.2 | 18.2 | — | 7.3 |
| Material purchasing | 51 | 87.3 | 19.6 | 7.8 | 54.9 | 31.4 | 3.9 | 3.9 | — | 7.8 |
| Chemical products | 41 | 82.9 | 34.1 | 14.6 | 53.7 | 41.5 | — | 7.3 | — | — |
| Others | 23 | 100.0 | 26.1 | 17.4 | 65.2 | 52.2 | 13.0 | 4.3 | — | — |
| Design centers and R&D bases | 32 | 59.4 | 65.6 | 31.3 | 56.3 | 37.5 | 12.5 | 18.8 | 6.3 | 9.4 |
| Total[d] | 295 | 85.1 | 31.9 | 13.9 | 60.7 | 40.7 | 10.2 | 13.9 | 2.0 | 5.4 |

*Source:* JETRO 1993, table V.2.

*Note:* Reasons for localization were categorized as follows:

A. It is necessary to supply products meeting local needs.
B. It is necessary to employ foreign researchers and use their ideas and conceptions in order to expand the scope of R&D.
C. It is necessary to promote joint research projects with foreign companies and research centers.
D. It is necessary to know local technological trends and adapt to the greater technological competition as soon as possible.
E. It is necessary to reduce the lead time from R&D to commercialization.
F. There is a shortage of researchers and developers in Japan.
G. The R&D localization is indispensable as one of the approaches to have inside information in the European R&D sector.
H. The R&D localization has been strongly desired by local governments and other authorities.
I. Others.

[a] Italy, Spain, Portugal, and Greece.

[b] Finland, Norway, Sweden, Denmark, and Ireland.

[c] Austria, Switzerland, and Iceland.

[d] Some of the 276 respondents were diversified and are reported in two sectors.

**TABLE 4.12. Activities Carried Out by Japanese Design Centers and R&D Facilities in Europe, by Industry**

| | Total Replies | Type of Activity (in percentage)[a] | | | | | |
|---|---|---|---|---|---|---|---|
| | | A | B | C | D | E | Total |
| Chemicals | 28 | 17.9 | 35.7 | 25.0 | 21.4 | — | 100.0 |
| Rubber products | 8 | 25.0 | 25.0 | 25.0 | 25.0 | — | 100.0 |
| Metal products | 9 | 11.1 | 33.3 | 44.4 | 11.1 | — | 100.0 |
| General machinery | 24 | 4.2 | 33.3 | 45.8 | 8.3 | 8.3 | 100.0 |
| Electronic and electrical appliances | 38 | 2.6 | 28.9 | 42.1 | 18.4 | 7.9 | 100.0 |
| Electronic parts | 22 | — | 31.8 | 40.9 | 22.7 | 4.5 | 100.0 |
| Transport equipment | 12 | — | 33.3 | 50.0 | 16.7 | — | 100.0 |
| Transport equipment parts | 19 | 15.8 | 36.8 | 26.3 | 10.5 | 10.5 | 100.0 |
| Precision machinery | 8 | 12.5 | 25.0 | 50.0 | 12.5 | — | 100.0 |
| Design centers and R&D facilities[b] | 15 | 26.7 | 33.3 | 20.0 | — | 20.0 | 100.0 |
| Other | 42 | 11.9 | 45.2 | 19.0 | 19.0 | 4.8 | 100.0 |
| Total | 225 | 10.2 | 34.7 | 33.3 | 16.0 | 5.8 | 100.0 |

*Source:* JETRO 1991, summary table 30.

*Note:* One hundred five companies supplied information on their type of research. Since many did more than one type, there were 204 replies. Since some of the respondents were allocated to two industries, their replies were included in both, thus the total number of answers reported in the table is 225.

[a]Activities were categorized as follows:

A. Basic research.
B. Product development.
C. Change of product design and specifications.
D. Development of production processes and technologies.
E. Other.

[b]Independent facilities not associated with a particular subsidiary.

meeting the needs of the European market still take the predominant positions. Thus 70 (66.7 percent) of the 105 respondents included product development as part of their work, while 32 (30.5 percent) included development of production processes and technologies (see footnote to table 4.12 regarding number of replies). Change of product design and specification, the SL adaptation role traditionally considered the primary function of overseas R&D, retains a position alongside the more contemporary LIL development role with 67 (62.8 percent) of the 105 respondents including this work. Basic research was carried out in 22 (21.0 percent) of the responding facilities and was, as would be expected, relatively more prevalent in the independent, IIL type, design centers and R&D facilities than in those associated with production units and therefore attributed to particular industries in the table.

CHAPTER 5

# Japanese R&D Laboratories in the United Kingdom

The roles and characteristics of Japanese MNEs' laboratories in the United Kingdom were investigated in a survey carried out by the authors in 1992–93. Using an address list supplied by JETRO London, we surveyed the 48 Japanese R&D facilities that it covered. This provided 19 usable replies. The responding laboratories operated in a wide range of industries (notably including electronics, chemicals, pharmaceuticals, plastics, and automobiles), which meant that there tended to be too few respondents in individual industries to allow classification of results by sector. However, another subdivision into three groups did prove to be analytically rewarding. Here group 1 comprises 11 R&D units whose primary role is clearly that of providing direct support to a United Kingdom-based Japanese production operation. Group 2 is made up of 5 laboratories that are essentially independent of any other United Kingdom-based operations belonging to their Japanese parent company. The 3 laboratories that constitute group 3 are, like those in group 2, stand-alone units without a substantial and primary integration into other United Kingdom-based operations. These group 3 units did, however, provide some types of support to a specific U.K. operation, without this being their central motivation as was the case for the group 1 labs.

## Characteristics of the Laboratories

Overall 13 of the 19 Japancse laboratories were set up in 1988 or subsequently, these comprising 7 of the 8 (88 percent) of the mainly independent (groups 2 and 3) units and 6 of the 11 (55 percent) that are closely associated with production operations (group 1). This fact may provide broad support for the suggestion that there has been some movement toward the desire to incorporate IIL-style work in the U.K. labs, after an initial preference for the SL- or LIL-type operations that are directly associated with production subsidiaries. Such a shift would clearly be in line with our view of a widening of the roles of overseas R&D of MNEs. Of the responding laboratories, 15 gave dates for the establishment of an associated production unit.[1] Five of these were set up in 1988 or subsequently, while 7 had been created prior to 1980. Of the 7 production facilities set up before 1980, only 3 acquired its associated laboratory within 5 years of its own establishment, while all except one of those set up since 1980 acquired its R&D unit within two years. This finding is compatible with the view that the competitive environment of overseas production subsidiaries is increasingly such that they need the close support of their own technology operations.

The majority of the labs were established as fresh installations, with this being rather more prevalent in the case of independent units (7 of the 8) than those associated with production facilities (6 of the 11). Three of the associated laboratories had been acquired as part of a company involved in a merger or takeover, while the remaining independent lab had also entered the Japanese MNE in this way before securing its more autonomous role. One laboratory originally joined its Japanese group as the result of the acquisition of an existing independent R&D facility but then became primarily associated with a United Kingdom-based production affiliate. The remaining lab was set up as part of a collaboration in a joint R&D venture with another company.

All except one of the laboratories reported the number of their employees. The size range varied from 4 labs (all in group 1) that had fewer than 10 employees, through 10 with from 10 to 39, to 2 with 40, another lab with 47, and the clear largest with 114. The average number of employees was 27 per lab, with this number higher in independent labs (38; boosted by the inclusion of the largest) than in the associated facilities (20). Another question relating to employment patterns in these Japanese MNE laboratories in the United Kingdom showed that local scientific personnel usually predominated. In the 8 group 1 units that answered, 4 had more than 90 percent U.K. personnel, though the other 4 had less than 75 percent. In the 8 group 2 and 3 labs, 5 reported more than 90 percent local personnel and none less than 75 percent. The majority of the remaining personnel usually came from the Japanese parent company, with 2 respondents also indicating that they secured personnel from other labs of the Japanese group and only 2 reporting recruitment elsewhere. The modestly greater tendency to commit Japanese personnel to group 1 labs may be related to the need to transmit elements of the parent company's current technology to these units, in order that they may then develop it effectively for the European markets to be supported by their allied production operations. In groups 2 and 3, where the quality of the U.K.

1. This number includes one group 2 lab. However, other aspects of the reply indicate no systematic operational links with the production unit, so the association may be nominal and administrative with no operational content.

scientists is a key factor, the need for mediation of Japanese personnel may be seen as less important.

### Role of the Laboratories

In classifying laboratories to the three groups, careful attention was paid to their replies to several of the questions. However, central to the process was a question that asked them to grade three types of work as (1) their only role, (2) part of their role, or (3) not a part of their role.[2] To directly support local (i.e., United Kingdom) production or marketing operations of the Japanese MNE emerged as the only role of 2 group 1 labs, part of the role of 8 others, and not a role for the other in this group. It was also part of the role of all 3 group 3 labs and in addition reported as a secondary commitment of 2 of the essentially independent group 2 units.

The second suggested role for the U.K. labs was to support production or marketing operations of the Japanese MNE in another country. The one group 1 lab that said it did not encompass the first role in fact specialized entirely in this function, while 6 of the other units in this group rated it as part of their role, though 4 excluded it. It was also part of the role of 2 group 3 labs and again a secondary occupation of 1 in group 2. The final suggested role was to operate as an independent research facility, unattached to other local operations of the MNE. Two of the group 2 laboratories assessed this role as their only function, while the other 2 respondents in group 2 and all 3 in group 3 considered it a part of their role. Three of the group 1 labs considered that part of their work could be evaluated as fulfilling this role, but 8 of them totally excluded it.

### Associated Laboratories

A number of questions were directed explicitly to laboratories that felt that at least part of their role was to support a production subsidiary. Of the 14 relevant (group 1 and 3) cases, 13 replied to a question on their associated subsidiary's market orientation. Only one of these subsidiaries was felt to focus entirely on the U.K. market, with the 12 others also including both other European Union and non-European Union European countries in their markets and 11 of them also supplying markets outside Europe. Another question asked these labs to assess the main objective of their associated production subsidiary by checking off what they considered the most relevant of three suggested alternatives. This question was answered by all group 1

---

2. One laboratory did not reply to this question and was classified to group 2 on the basis of other parts of its reply.

cases (though 3 checked 2 alternatives) and 2 group 3 cases. The more traditional role of supplying established products emerged as still the most prevalent, being checked by 9 group 1 cases (3 in combination with another role) and 1 from group 3. To create and supply products that are additions to the MNE group's established range was checked by 4 group 1 cases (2 in combination with the previous role) and the other group 3 respondent. The role of supplying intermediate products (e.g., component parts) to other parts of the MNE group did not emerge as a notable role, being checked by only 1 respondent and then in combination with the first role.

The next question directed to these laboratories investigated the nature of their own work. Thus they were asked to evaluate three types of work as being either (1) their most important work, (2) part of their work, or (3) not a part of their work. By allocating replies of most important a value of 3, part of their work a value of 2, and not a part of their work a value of 1, average responses (AR) were calculated to summarize the results. The first role for the labs was the traditional SL-style one of adapting existing products. This role received an AR of 2.09 for the group 1 labs and 1.33 for those in group 3. The complementary role of adapting existing production processes was slightly less common, with ARs of 1.82 and 1.00 for the group 1 and 3 cases, respectively. The more contemporary LIL-oriented role of creation of new products received an AR of 2.09 from the group 1 respondents and 2.00 from those in group 3. This result clearly places product creation alongside product adaptation among these labs' main functions, suggesting that the previously noted predominance of the supply of established products as the main role of their associated producing subsidiaries hides an emerging position for the supply of new items created for European markets. Though the small number of cases precludes strong conclusions, two other brief observations may be pertinent. The fact that the group 3 labs provide lower ARs for all three of these roles reflects the fact that their association with production units is not their primary role, as it is for group 1 cases, and that they also have a commitment to other independent-laboratory functions. Second, contributing to the creation of new products is distinctively the strongest of these roles for group 3 labs, in a way that is not the case for group 1 facilities. Thus the higher scientific abilities of the group 3 cases, reflecting their essentially independent status, seems to be called upon to assist in product development, while it is not as often considered necessary for product adaptation.

Another question aimed at the group 1 and 3 cases in the survey of Japanese labs in the United Kingdom asked those that felt that their work helped create new products that would then be supplied by an associated production facility to evaluate three reasons for undertaking this role as either (1) very important, (2) quite important, or (3) not important. Overall, 9 group 1 labs and all 3 in group 3 felt that their work qualified them to answer this question. Two reasons were rated as especially important influences for taking part in product creation. First, such a role clearly aims "to encourage the competitiveness of the subsidiary in the European market," with the group 1 cases providing an AR[3] of 2.66 and those in group 3 an AR of 2.33. Second, the suggestion that the product development role could be seen "as part of the Japanese parent company's strategy for a globalized approach to innovation" clearly strongly complements the first perspective, achieving an AR of 2.55 from group 1 and 3.00 from group 3 respondents. Again with the proviso of the small number of replies, it is interesting to observe the reversal of emphasis of the two groups' responses to these factors. Thus the more independent group 3 cases encompass rather more strongly the globalized perspective, while the much more directly production-linked group 1 units acknowledge rather more precisely the dictates of the subsidiaries' European competitive environment.

The third factor addressed in this question, "to make full use of scientific abilities in this country," was also quite highly valued, with ARs of 2.00 and 2.33 from the group 1 and 3 labs, respectively. This response indicates that whereas the need to perform product creation for the European market, as part of the global-innovation strategy, is the paramount motivation, the decision to use U.K. facilities to play this role has also been extensively influenced by the supply-side-related ability of the U.K. units' capacity to perform it effectively.

## Independent Laboratories

Some other aspects of the R&D of Japanese companies in Europe were investigated in the survey of their laboratories in the United Kingdom through three questions addressed to the independent or stand-alone units. In addition to the 8 group 2 and 3 facilities, these questions were answered by 2 from group 1, which felt that they also included some aspects of independent labs' work in their activities. The first of these questions asked respondents to evaluate as (1) their most important work, (2) part of their work, or (3) not part of their work, three elements of R&D. The most prevalent of the three types of work was basic scientific research, with an AR of 2.20,[4] followed by the development of new product concepts with 2.00. Assisting the implementation of critical changes to existing products (adaptation) was clearly the least prevalent type of work in these laboratories, with an AR of 1.60. There were no significant differences in these replies by grouping of labs. Overall these results indicate that these independent R&D facilities tend to contribute more

3. With very important given the value of 3, quite important 2, and not important 1.

4. With most important given the value of 3, part of their work given the value 2, and not a part the value 1.

to the first phase of a globalized approach to innovation, by generating new precompetitive technology and the outlines of a new product concept, than to its second phase, application to the specific European-market context.

Where R&D facilities are not tied strongly to the support of operations targeted at the supply of particular national markets, their location may be more likely to respond to supply-side factors (availability, quality, and cost of relevant inputs). Thus another question answered by these 10 laboratories asked them to evaluate as (1) a major influence, (2) of some influence, or (3) of no influence, four possible factors relating to the decision to set up the facility in the United Kingdom. The first factor, "availability of talented scientists," received an AR[5] of 2.60 from the 5 fully independent group 2 respondents and 3.00 from the 5 group 1 and 3 respondents with some degree of association with a production operation. Next, "the country has a strong tradition in our areas of science" received an AR of 3.00 from the group 2 labs and 2.40 from those in groups 1 and 3. The differences within these replies may suggest that the associated labs, perhaps helping production units apply knowledge in an ambitious way, may need generally high-quality scientists who can assimilate and use established technology, while the independent ones, more clearly oriented to contributing to new research results, may be more motivated by inputs reflecting specialized areas of science that represent the United Kingdom's particular tradition. The other two influences investigated emerged as much less relevant, with the presence of "an accessible university system for international recruitment" getting an AR of only 1.60 and "the presence in the country of similar labs of our competitors" only 1.20.

Further indications of the role and orientation of these stand-alone units were tested in the third question, which asked whether or not five sources of project ideas were relevant to their programs of work. All of the 9 that answered the question felt that ideas originating "within our own lab" were relevant, with 3 of these not considering any of the other sources applicable. Four indicated that ideas "from the parent company's labs in Japan" were implemented in their projects, but only 2 felt it applied to ideas "from other overseas labs of the Japanese parent company." While 4 of the laboratories considered that relevant project ideas came "from other host-country scientific institutions (e.g., universities)," only 2 felt such ideas emerge "as part of collaborative work with labs from other companies."

## Laboratories' Scientific Links

It is a clear theme of this book that overseas subsidiaries of MNEs play an interactive role in the creative activity of the overall group. Thus a globally

5. With major influence given the value of 3, some influence the value 2, and no influence the value 1.

competing MNE is likely to have a pervasive dominant technological trajectory, which is likely to influence the evolution in the operations of each of its subsidiaries. These subsidiaries, however, often pick up this underlying technology and apply their own resources (managerial and marketing, as well as scientific) to it in order to implement the second phase of a global-innovation strategy as competitively as possible in their own market area. Though this practice clearly implies a very strong demand-side (i.e., market-based) influence on the decision to set up R&D facilities targeted at a local market, we have also seen that where in a large regional market, such as Europe, such a facility is located may depend to some extent on supply-side factors, notably the quality of the local science base. Where these R&D subsidiaries play a role in the supply of basic research to the precompetitive first phase of an MNE's global-innovation program, availability of quality scientific inputs is likely to be an even more influential factor. Thus alongside their integration into the wider operations of their MNE, these R&D labs are also likely to interact significantly with other elements of the scientific community in their host country.

The survey of Japanese R&D labs in the United Kingdom investigated such collaborative links with the local science base, asking respondents to evaluate their associations with four types of institution as extensive, moderate, or nonexistent. Overall these links were strongest with universities, where the group 2 (independent) labs provided an AR[6] of 2.60, those in group 3 (semi-independent) one of 2.00, and those in group 1 (closely associated with the work of a production facility) 1.55. These results are clearly in line with an expectation that university links would be more likely to be expected to be fruitful for the more basic-research precompetitive orientation of independent laboratories than for the mainly development aims of the associated facilities. Links with independent U.K. research labs, which (like university labs) are expected to encompass expertise in specialized areas of mainly precompetitive work, were rated at an AR of 2.00 for group 2 cases and 1.36 and 1.33 for groups 1 and 3, respectively. Thus, while less prevalent, these links seem to reflect the same motivation as those with universities.

The third type of link was with industry-funded research laboratories. These may differ from the two previous forms of collaborative institution by having a much clearer orientation toward, and understanding of, the commercial as well as the technological environment of a particular industry. Though links with such labs are not yet well established, the pattern does tend to reflect these characteristics. Thus the group 2 labs (which predominated in the two previous types of collaboration) reported no such links, while those in groups 1 and 3 provided ARs of 1.36 and 1.67, respectively. For collaborative

6. With extensive given the value of 3, moderate 2, and nonexistent 1.

work with local U.K. firms, the Japanese labs in group 1 provided an AR of 1.55, but such links for groups 2 and 3 cases were limited to 1 "moderate" association. It may, therefore, be that such interfirm collaborations as existed were mostly concerned with the commercial application of established knowledge.

The types of collaborations just described may benefit the institutions in the U.K. science base by extending their research environment and thereby expanding their scope and range of experience. Another potential source of valuable learning that may be accessible to host-country scientists in MNEs could be secondments to other labs in the group. So far this appears to have been relatively rare in the case of the Japanese labs in the United Kingdom, though it was suggested that it may become a more common practice when the newer units are more settled. Of the 18 labs that responded to a question in personnel mobility, none indicated that they moved U.K. researchers to parent labs in Japan frequently, only 7 did so occasionally, and 11 never did. Similarly, none moved personnel to other non-United Kingdom labs in the MNE group frequently and only 3 did so occasionally. Secondments to other U.K. scientific institutions (e.g., universities, industry research labs), as an additional means of tapping into U.K. basic science, was only a little more prevalent than intragroup mobility. Thus only 2 of the 18 respondents said this practice was frequent and 3 that it happened occasionally.

Another way that Japanese labs could spread benefits into the U.K. economy is through the provision of advice to U.K. companies that supply their associated Japanese production operations. When asked to evaluate the extent to which they offered such advice, 2 group 1 respondents said frequently, 7 sometimes, 1 rarely, and 1 never. Five of the group 2 and 3 labs rated this function as not relevant to the nature of their work, 2 said they never did it (though presumably considering it within their scope), and 1 did offer such advice, but only rarely. Ten of these respondents then answered a supplementary question that asked them to grade as (1) very important, (2) quite important, or (3) not important, three types of advice they could offer to U.K. suppliers. Advice on the specifications of the components or inputs to be supplied was predominant with an AR[7] of 2.90, but details of the production process used to supply inputs notably less so with an AR of only 1.80. Also quite rare (an AR of 1.60) was the provision of wider advice on industrial production processes. This finding suggests that the most commonly supplied advice is mainly related to the specific details of the supplier's role in support of the Japanese company's production operations, with broader knowledge that could also be beneficially applied to the suppliers wider operations relatively rare.

7. With very important given the value of 3, quite important 2, and not important 1.

## Laboratories' Views on Policies

A final question sought the responding Japanese laboratories' views on policy issues relating to their operations. Thus they were invited to indicate if they believed (1) very strongly, (2) quite strongly, or (3) not at all, in four suggested statements. The first statement was that "host-country policies influenced the decision to set up this R&D facility." Only 3 respondents held this view very strongly and 6 rejected it, to provide an AR[8] of 1.84. The second view related to the likelihood of the direction of the evolution of labs being subject to influence by host-country policy. Generally this was rated very similarly to the conditions of the laboratory's creation, with 3 respondents believing very strongly that "host-country policies influence decisions with regard to the development of this facility" and 7 rejecting that view (an AR of 1.79). No respondents believed very strongly that "host-country policies overemphasize support for basic research," and only 3 (all group 1) labs supported the view quite strongly. However, there was rather more support for the view that "host-country policies should provide more support for product development and innovation." As might be expected, this view was most widely held in the group 1 labs, where 4 of the 11 endorsed it very strongly and only 1 rejected it (an AR of 2.27). Only 1 of the group 2 and 3 labs supported this view strongly and 3 rejected it (an AR of 1.75).

8. With very strongly given the value of 3, quite strongly 2, and not at all 1.

CHAPTER 6

# Japanese Electronics MNEs' Subsidiaries in Europe

In this chapter we analyze aspects of the technological behavior of Japanese electronics subsidiaries operating in Europe, using results from a survey analysis carried out by Marina Papanastassiou in 1992–93. The full survey covered 145[1] manufacturing subsidiaries, of which 24 belonged to Japanese parent companies, 71 had U.S. parents, and 46 were subsidiaries of a European MNE operating in a country other than its home country.[2] Of the 24 Japanese respondents, 15 were in electronics-related (including telecommunications) industries. Since reporting of the full Japanese sample would therefore strongly reflect that industry, and also because of the particular interest in the notable success of Japanese firms in Europe in the industry, we focus analysis on these 15 electronics subsidiaries, reporting the other 9 separately for purposes of comparison. Of the U.S. and European subsidiaries in the survey, 18 were in the electronics industry, and these are also reported to add an extra comparative dimension.[3]

## Market Orientation and Roles of Subsidiaries

As the discussion in chapter 2 has emphasized, the extent and nature of technological activity in MNE subsidiaries depends very strongly on their role, which is in turn closely associated with their market orientation. This section therefore looks at these aspects of the three groups of subsidiaries that are being subjected to analysis. In terms of general market orientation, table 6.1 shows that the Japanese electronics subsidiaries were very strongly dedicated to the supply of export markets. Though this export orientation is

1. Questionnaires were sent to 313 subsidiaries in the United Kingdom, providing a response rate of 31.6 percent, and to 220 in Belgium, Greece, and Portugal, for a response rate of 20.9 percent.

2. The survey covers subsidiaries in 4 countries: the United Kingdom (99 subsidiaries), Belgium (20), Greece (16), and Portugal (10).

3. Further evidence on Japanese MNEs in Europe will become available shortly from a new survey carried out by the authors. It includes 69 Japanese subsidiaries among 190 replies from MNE affiliates in the manufacturing sector in the United Kingdom.

**TABLE 6.1. Export Orientation of Japanese Electronics Companies' Subsidiaries in Europe**

| | Proportion of Production Exported (percentage of respondents) | | | | | |
|---|---|---|---|---|---|---|
| | 0 | 0.1% to 10% | 10.1% to 25% | 25.1% to 50% | More than 50% | Total |
| Japanese electronics subsidiaries | 0.0 | 6.7 | 6.7 | 20.0 | 66.7 | 100.0 |
| Other Japanese manufacturing subsidiaries | 22.2 | 0.0 | 0.0 | 22.2 | 55.6 | 100.0 |
| Other electronics subsidiaries[a] | 30.8 | 15.4 | 7.7 | 23.1 | 23.1 | 100.0 |

*Source:* Marina Papanastassiou data base.
*Note:* Data covers subsidiaries in the United Kingdom, Belgium, Greece, and Portugal.
[a]Subsidiaries of U.S. or European MNEs.

somewhat greater than that of the other Japanese subsidiaries, the clearest contrast is with the much more host-market-focused U.S. and European electronics subsidiaries.[4] The European market is clearly central to the strategic positioning of these Japanese electronics subsidiaries, with both the EU and other European countries being part of the target markets for the exports of 14 of the 15, though all except 4 of them also supplied markets outside of Europe.[5]

Another potentially important market-orientation characteristic of subsidiaries is the extent to which they supply goods to other parts (either production or marketing operations) of their MNE group. Such intragroup exports are shown in table 6.2 to be quite strong among the Japanese electronics subsidiaries, where this aspect of supply networking is clearly much better established than in the other Japanese subsidiaries. In this aspect of trading behavior, it is the other electronics subsidiaries that most closely resemble the Japanese operations in this industry, with table 6.2 showing that of the exports that these subsidiaries do undertake, the intragroup component is quite strong.

The third aspect of exporting investigated was the position of intermediate products (e.g., component parts or goods requiring further processing), as shown in table 6.3. Despite their substantial commitment to intragroup trade, this type of good played a limited role in the exports of the Japanese electronics subsidiaries. This finding suggests that these subsidiaries were specializing in the supply of particular finished goods to an integrated Eu-

4. For discussion of the host-market focus of European-MNE subsidiaries in particular, see Papanastassiou and Pearce 1993, 1994c.

5. For further data on the market orientation of Japanese electronics subsidiaries in Europe, see Suzuki 1994.

**TABLE 6.2. Intragroup Exports in the Trade of Japanese Electronics Companies' Subsidiaries in Europe**

| | Intragroup Exports as a Proportion of Total Exports (percentage of respondents) | | | | |
|---|---|---|---|---|---|
| | 0 | 0.1% to 10% | 10.1% to 50% | Over 50% | Total |
| Japanese electronics subsidiaries | 28.6 | 14.3 | 21.4 | 35.7 | 100.0 |
| Other Japanese manufacturing subsidiaries | 62.5 | 25.0 | 0.0 | 12.5 | 100.0 |
| Other electronics subsidiaries[a] | 44.4 | 11.1 | 11.1 | 33.3 | 100.0 |

*Source:* Marina Papanastassiou data base.
*Note:* Respondents were asked what percentage of their exports went to other parts of their MNE group. Data covers subsidiaries in the United Kingdom, Belgium, Greece, and Portugal.
[a]Subsidiaries of U.S. or European MNEs.

ropewide marketing network of the MNE. This specialized position can have been either *allocated* to the subsidiary in the form of RP supply of an established good or *earned* in the form of the authorization to create and supply a new good as an RPM operation. The other electronics subsidiaries also found little room for intermediate product supply, suggesting that, within the context of their generally lower export orientation, their relative commitment to intragroup exports may again reflect this type of final-product-supply networking. In the other Japanese subsidiaries a much greater presence of intermediate-goods exports combines with a limited degree of intragroup trade. Thus Japanese MNEs that themselves specialize in the supply of component parts (notably in the automobile industry) have set up European operations to continue to service their established customers.

This evidence on market orientation has clearly suggested a distinctive

**TABLE 6.3. Intermediate-product Exports in the Trade of Japanese Electronics Companies' Subsidiaries in Europe**

| | Intermediate Product Exports as a Proportion of Total Exports (percentage of respondents) | | | |
|---|---|---|---|---|
| | 0 | 0.1% to 50% | Over 50% | Total |
| Japanese electronics subsidiaries | 69.2 | 7.7 | 23.1 | 100.0 |
| Other Japanese manufacturing subsidiaries | 12.5 | 12.5 | 75.0 | 100.0 |
| Other electronics subsidiaries[a] | 88.9 | 0.0 | 11.1 | 100.0 |

*Source:* Marina Papanastassiou data base.
*Note:* Respondents were asked what percentage of their total exports were intermediate goods (e.g., component parts or goods requiring further processing). Data covers subsidiaries in the United Kingdom, Belgium, Greece, and Portugal.
[a]Subsidiaries of U.S. or European MNEs.

strategic approach to the supply of the European market by Japanese electronics MNEs. Further evidence on the position taken by individual subsidiaries in the implementation of this strategy can be obtained by a question in the survey that asked respondents to indicate which of three suggested roles they considered their main one. Though over half the Japanese electronics respondents (table 6.4) felt that the traditional role of supplying an established product remained their main one, it was much less pervasive for them than for either other Japanese subsidiaries or for non-Japanese electronics subsidiaries.

The more contemporary (RPM or WPM) subsidiary role of creating and supplying a product that is a new addition to an MNE group's range has already emerged rather more strongly in Japanese electronics subsidiaries than in those from other countries' MNEs and is also clearly more pervasive in Japanese electronics operations than in those in the other manufacturing industries. Recognition of the intensity of global competition in the industry and a particular willingness, or perceived need, to respond to specific European conditions with new products by Japanese MNEs may both contribute to their relatively strong support for creative subsidiaries within their European operations.

The third role defined in the questionnaire was "to supply an intermediate or final product to other parts of the MNE group." That this role emerges as somewhat more prevalent in Japanese electronics subsidiaries than in either of the other two groups covered is in line with the patterns indicated by the analysis of the export data. Indeed, while relatively few respondents were inclined to see this predominantly networked role as their main one, it seems

**TABLE 6.4. Roles of Japanese Electronics Companies' Subsidiaries in Europe**

| | Percentage of Respondents | | |
|---|---|---|---|
| Roles[a] | Japanese Electronics Subsidiaries | Other Japanese Manufacturing Subsidiaries | Other Electronic Subsidiaries[b] |
| To supply an established product | 58.8 | 81.8 | 76.9 |
| To create and supply a new product in the MNE group | 23.5 | 9.1 | 15.4 |
| To supply an intermediate or final product to other parts of the MNE group | 17.6 | 9.1 | 7.7 |
| Total | 100.0 | 100.0 | 100.0 |

*Source:* Marina Papanastassiou data base.
*Note:* Data covers subsidiaries in the United Kingdom, Belgium, Greece, and Portugal.
[a]Respondents were asked to check off the role they considered their main one.
[b]Subsidiaries of U.S. or European MNEs.

likely that many of the others (again especially the Japanese electronics subsidiaries) include significant intragroup trade alongside their open-market arms-length supply role. Overall, then, there seems to be some suggestion that Japanese electronics subsidiaries in Europe have reacted rather more proactively to the competitive needs of the industry, with clearer evidence of the emergence of product-creating RPM operations alongside continued pursuit of cost-efficient production through RP subsidiaries.

## Sources of Technology in Subsidiaries

Respondents to the survey were asked to evaluate the relative importance of three sources of technology that might be used in their operations. The first of these sources was technology imported from elsewhere in their MNE group. Such group technology can take two forms when adopted by a subsidiary. First, it can be the knowledge already embodied in an established product that is to be produced (perhaps with some adaptation) by the subsidiary. Second, it may be group-level technology (i.e., that underpinning a new product concept) that is then to be developed into a distinctive product to meet the specific needs of a particular (i.e., European) market. The results in table 6.5 show that this imported technology from elsewhere in the group is the predominant source of the technology used in the operations of Japanese subsidiaries (in electronics and other manufacturing industries) in Europe, indicating that these companies have substantial confidence in the commercial viability of

**TABLE 6.5. Sources of Technology Used in Japanese Electronics Subsidiaries in Europe**

| | Source of Technology (average response)[a] | | |
|---|---|---|---|
| | A | B | C |
| Japanese electronics subsidiaries | 2.71 | 1.46 | 1.64 |
| Other Japanese manufacturing subsidiaries | 2.75 | 1.75 | 2.00 |
| Other electronics subsidiaries[b] | 2.27 | 1.44 | 2.22 |

*Source:* Marina Papanastassiou data base.

*Note:* Data covers subsidiaries in the United Kingdom, Belgium, Greece, and Portugal.

[a]Sources of technology were categorized as follows:

A. imported technology from elsewhere in the group.

B. established host-country technology.

C. results of R&D carried out by the subsidiary.

Respondents were asked to grade each source as (1) the main source, (2) a secondary source, or (3) not a source. The average response was calculated by allocating a value of 3 to the main source, 2 to a secondary source, or 1 to not a source.

[b]Subsidiaries of U.S. or European MNEs.

their existing knowledge at these levels of its development. The table also suggests that such confidence in their established technology is less clearly the case for the non-Japanese electronics respondents.

Neither the Japanese nor the other electronics subsidiaries seem to find much to interest them in established host-country technology as a source of knowledge to support their operations, though Japanese subsidiaries in the other manufacturing industries do make somewhat greater use of it. The third source of technology analyzed was the results of R&D carried out by the subsidiary. It emerged most strongly as a relevant source in the non-Japanese electronics subsidiaries, where it may have emerged into a quite strong position in reflection of a relative lack of confidence in (or availability of) importable group technology. By contrast, in-house R&D as a source of technology was least prevalent in the Japanese electronics subsidiaries. Though confidence in the adequacy of imported group technology as a sufficient source could be an explanation for the lack of their own R&D in this case, it seems unlikely to be an adequate one given the suggestion of the presence of a product development role in many of these subsidiaries and the competitive pressures for decentralized technological dynamism in the industry. A more plausible explanation for the low level of in-house R&D in these Japanese electronics subsidiaries reflects another characteristic suggested by earlier discussion in this chapter. That is the fact that many of them appear to play specialized roles within a carefully articulated and balanced European supply network of their MNE group. Where such a network exists, it may have been decided to provide it with centralized R&D support from a major European laboratory, from which even product-development subsidiaries can commission any scientific work they require.[6]

## Roles of Subsidiary R&D Laboratories

Another question in the survey asked those subsidiaries that had R&D labs to evaluate the relative importance of three types of work that they might do. In fact, 5 of the 15 Japanese electronics subsidiaries replied to this question, compared with 4 of the 9 Japanese subsidiaries in other industries and 10 of the 18 U.S. and European electronics subsidiaries. The first two roles covered in table 6.6 relate to the ways in which subsidiary-level R&D could be applied to imported group-level technology in its two possible (embodied and disembodied) forms. Clearly the three groups of subsidiaries were somewhat more likely to use their laboratories to develop a new product than to adapt an existing product or production process. This result may indicate that subsid-

6. Preliminary results from the authors' new survey provide additional support for this perspective. See Papanastassiou and Pearce 1994d.

**TABLE 6.6 Japanese Electronics MNEs' Subsidiaries' Evaluation of the Importance of Various Types of Work in their Laboratories in Europe**

| | Type of Research (average response)[a] | | |
|---|---|---|---|
| | A | B | C |
| Japanese electronic subsidiaries | 2.40 | 3.00 | 1.40 |
| Other Japanese manufacturing subsidiaries | 2.00 | 3.00 | 1.75 |
| Other electronics subsidiaries[b] | 2.00 | 2.75 | 1.75 |

*Source:* Marina Papanastassiou data base.
*Note:* Data covers subsidiaries in the United Kingdom, Belgium, Greece, and Portugal.
[a]Types of research were categorized as follows:
A. adaptation of the product or production process.
B. development of a new product.
C. provision of scientific knowledge to a broader research project by the MNE group.
Respondents were asked to evaluate each type of work as either important, relatively important, or not important. The average response was then calculated by allocating responses of important the value 3, relatively important 2, and not important 1.
[b]Subsidiaries of U.S. or European MNEs.

iaries that incorporate laboratories have a pervasive tendency to use them mainly as LILs, probably supporting RPM operations in order to develop distinctive European product variants from technology imported in the form of the outlines of a major new product concept. The IIL-style role of provision of scientific knowledge to a broader research project of the MNE group was much less well established in the R&D operations of the responding subsidiaries. Though the small difference and limited coverage preclude a strong conclusion, the lower response to this third role in Japanese electronics subsidiary labs is compatible with the view that in this industry the Japanese MNEs are particularly confident that the home-country technological capability can continue to take responsibility for the background basic technology necessary to sustain their leading competitive position.

## Links with Local Scientific Institutions

As we noted in the previous chapter, one way in which subsidiary laboratories may seek to extend the technological scope of their MNE group is through collaborative research with host-country scientific institutions. At the basic or precompetitive end of the R&D spectrum, collaborations with universities or independent research laboratories may allow MNEs to tap into the more important long-term speculative work, getting an early sense of, and involvement with, possible sources of revolutionary future product generations. Japanese electronics subsidiaries are shown in table 6.7 to feel less need for this

**TABLE 6.7. Japanese Electronics MNEs' Subsidiaries' Evaluation of their Collaborative Research with Local Institutions in Europe**

| | Research Collaboration (average response)[a] | | | |
|---|---|---|---|---|
| | Universities | Independent Research Laboratories | Industry Research Laboratories | Other Firms |
| Japanese electronics subsidiaries | 1.36 | 1.00 | 1.18 | 1.40 |
| Other Japanese manufacturing subsidiaries | 1.50 | 1.38 | 1.25 | 1.25 |
| Other electronics subsidiaries[b] | 1.55 | 1.10 | 1.27 | 1.55 |

*Source:* Marina Papanastassiou data base.

*Note:* Data covers subsidiaries in the United Kingdom, Belgium, Greece, and Portugal.

[a]Respondents were asked to evaluate each type of collaboration as extensive, moderate, or nonexistent. The average response was then calculated by allocating responses of extensive the value of 3, moderate 2, and nonexistent 1.

[b]Subsidiaries of U.S. or European MNEs.

extension of their research scope than either other Japanese subsidiaries or electronics subsidiaries from the United States and Europe. This finding again is compatible with a continued confidence in the home-country technological sources that generated the existing competitiveness of these companies.

Industry research laboratories may move closer to the applied research and development end of the spectrum, incorporating a clearer perspective on the technological needs of their particular industry. However, few of the respondents seemed to need this type of collaboration to help improve their ability to apply their technology commercially in Europe.

Collaborations with other firms could take a horizontal form, with both partners contributing at more-or-less the same point on the R&D spectrum. Collaborations in precompetitive basic research would perhaps be the most likely ventures of this type. Vertical interfirm collaborations could involve one firm providing expertise in commercializing knowledge effectively in a particular market and the other deriving and supplying the high-quality knowledge itself. In light of the indication that Japanese electronics companies do not lack confidence in their underlying scientific capacity, it may well be that the collaborations with other firms indicated in table 6.7 involve the vertical type, with the other firms mainly providing knowledge of the needs of the European markets. The relative prevalence of interfirm collaborations of non-Japanese electronics companies may indicate the need for both horizontal and vertical research alliances, perhaps reflecting a perceived vulnerability vis-à-vis their Japanese competitors.

Overall the technological behavior of Japanese electronics MNEs in Europe seems to incorporate elements of the more contemporary strategic approaches. These approaches seek to build on a sense of confidence in their technological capacity rather than to reflect its vulnerability, but they do indicate a strong awareness that this technological strength needs to be applied to the European market in ways that fully respond to its distinctive needs.

CHAPTER 7

# Conclusions

It has been the aim of the earlier chapters of this book to document and evaluate the emergence of overseas R&D facilities in Japanese MNEs and to assess the motivations behind it. The manner in which such laboratories support the competitiveness of Japanese firms' operations in Europe has been emphasized. In this chapter we summarize the background to this new dimension in the pursuit of global competitiveness and evaluate its implications for the companies and for the countries in which the laboratories are located. One concern of the analysis has been to assess the role of host countries' science bases in decisions concerning decentralized R&D units in Japanese MNEs. The consequence of this assessment for the effective use and development of these national science bases emerges as an issue of considerable potential significance.

The various evidence discussed in the earlier chapters indicates that the predominant role played by the overseas R&D laboratories in Japanese MNEs remains to support the effective commercial application of the groups' current technology in particular markets. However, compared with an earlier situation in which it was generally believed that the role of such decentralized R&D was to help adapt established products for national markets as part of an import-substituting strategy, there appear to have been two key changes in the nature of this support role. First, the market targeted by the production and marketing subsidiaries with which the laboratories are associated has often extended from the national to the regional (e.g., European) as easier trade conditions allow individual subsidiaries to export to wider markets. Second, the laboratories work with production and marketing to develop products for those markets, taking full account of all relevant consumer taste and production conditions, rather than merely adapt established products in relatively marginal ways. This strategy, then, provides these laboratories and their associated subsidiaries with a role in the innovation process itself, ensuring that a vital new product concept reaches each major market in a fashion that is fully attuned to its distinctive needs and conditions.

These changes in the role of the laboratories in turn have implications for the factors influencing decisions regarding their location. It remains true that product development for a particular market is best carried out in facilities within the region. However, when this market itself is expanded from the national to the regional, the decision of where in the relevant wider area the laboratory is set up can be a matter of increased choice. Also, the enhanced ambition of the laboratories' work, from adaptation of embodied technology to development of new technology, is likely to make the quality of required scientific inputs much more important. Under these circumstances the extent and level of countries' abilities in particular areas of science is likely to become an increasingly important factor in determining the location of these development laboratories.

In order to be in a position for their overseas laboratories to assist in the competitive innovation of new product concepts, the MNE group needs first to derive the new background science underlying the vital new commercial ideas. As the pressures of global competition and innovation intensify, the range of technologies potentially relevant to the growth and evolution of firms widens. However, the international distribution of expertise in different areas of science is increasingly uneven, with the influence of agglomeration ensuring that individual countries establish and reinforce reputations for leadership in research in particular disciplines. Therefore, as firms seek high-quality inputs in a wider range of sciences and as the availability of these specialized inputs is perceived as increasingly focused on a limited range of locations, a strong argument for the establishment of an integrated network of decentralized laboratories, each working on a different aspect of basic precompetitive research in line with the capabilities of its host country, becomes increasingly clear. The evidence reviewed in this book suggests that this role for overseas R&D laboratories is certainly emerging in Japanese MNEs but at present seems to be mainly a secondary one, with product development still taking the predominant position. However, as the need for this global perspective on precompetitive research becomes clearer, and as the experience of the development laboratories demonstrates the viability of globally coordinated R&D efforts and increases knowledge of the scientific capabilities available in different locations, it seems likely that this role for overseas laboratories will increasingly complement the currently more established one.

It is, then, an overall conclusion of this book that at present the driving force behind most overseas laboratories in Japanese MNEs is what is demanded of them, in terms of supporting associated production and marketing operations. It is also argued, however, that the changing nature of this role and the emergence of the complementary decentralization of basic research will provide an increasingly strong influence for the ability of countries to supply relevant scientific inputs. It is out of the increasing intermeshing of the *global* technological activity of leading enterprises and *national* science bases that areas of debate concerning implications of the internationalization of MNEs' R&D are emerging.[1]

As a background to articulating these issues, let us assume that there are significant areas of potentially important scientific investigation that remain unlikely to be funded adequately by private enterprise, even very large globally competitive MNEs. Thus vital areas of basic (precompetitive) speculative research may be too expensive, involve too much risk and uncertainty, and provide results that may be difficult to appropriate adequately. If private enterprise will not risk funding these areas of research, it may be considered vital that some authority does support them, in the interest of the long-term capacity of science to enhance welfare and solve problems. Assume further that, for the present, the most likely sources of funding for precompetitive science emerge at the national level (or at most at the level of a regional group of countries) rather than at the global level. Then if, in the ways analyzed throughout the book, key users of these national science bases are MNEs (i.e., *nationally* created elements of science support the development of *international* firms that are owned and controlled abroad), three issues can be perceived to emerge.

First, does the intermeshing of national science bases and international firms' R&D programs have the potential to enhance the overall efficiency of science's contribution to global welfare? We can hypothesize ways in which it might do so. Thus it may be that if an MNE combines its own technological assets (its existing knowledge, its skill in the organization and commercialization of science, its current scientific objectives formed from comprehension of the wider technical and commercial context) with a country's science, then the quality and relevance of the work of that science base may be enhanced to levels that it would not have been possible for it to achieve in combination with local firms. Or, reversing the direction of argument, if an MNE is able to implement its program by collaborating with the most suitable selection of national sciences, then it may achieve more (i.e., make better use of its own existing assets) than if it constrained itself to the science base of its home country. The question is, then, whether or not the internationalization of R&D achieves this type of enhanced scientific effectiveness.

The second question concerns the distribution of the benefits that would accrue if the internationalization of MNEs' R&D and its intermeshing with national science communities did achieve the efficiency enhancing outcome just postulated. If this efficiency enhancement is attained through the comple-

1. Graham (1992, 201–6) discusses such issues from the point of view of Japanese MNEs' R&D in the United States. Pearce (1989, 191–96) and Cantwell (1993, 30–31) review the consequences of the networking of multinational R&D in MNEs.

mentary inputs of two independent parties (an MNE and the science base of a nation-state), it may be pertinent to ask if it then results in a *fair* distribution of benefits to these two partners. Given the intangible nature of both the inputs and the outputs of such a knowledge-related collaboration, it is massively difficult to define precisely a fair or equitable outcome in any particular case. What it is useful to attempt to achieve is some indication of the *nature* of the benefits the two parties might expect to achieve from participation in such a liaison and especially to point to particular circumstances in which host countries might be most likely to benefit or to be vulnerable to losses.

As a third question, however, it may be relevant to ask if it is really necessary to concern ourselves too much over the host-country consequences of foreign MNEs' R&D laboratories. Given the increasingly globalized operating perspectives of leading firms and the probable global viewpoint of most researchers (in the sense that they see themselves as primarily motivated to achieve progress in their areas of science, rather than as motivated to help science to benefit explicitly their own country), isn't it very narrow to ask how the MNEs' R&D can benefit individual host countries?

We hope that the answer to this question is, ideally, yes. Certainly the concerns of host countries should not reach the levels of a kind of research mercantilism. This situation could occur if a country were host to laboratories of foreign MNEs but were also the home country of MNEs that themselves possessed laboratories overseas. The government of such a country could try to maximize the benefits from foreign labs in the country and also those attained by its own MNEs' labs overseas in order to, in some sense, create a balance of research surplus. Such manipulation would be likely to severely damage the genuine welfare-creating potential of an internationalized firm-level approach to R&D.

Pragmatically, however, as long as precompetitive research is not going to be wholly or adequately funded by private companies, and as long as the infrastructure of basic research is predominantly created and supported at the national level, there may be reasons for fearing the results of countries becoming concerned at the use of their science base by MNEs.

There are two quite obvious and complementary reasons for this fear. First, if a government begins to suspect that a large share of the benefits of the scientific infrastructure it supports are attained in the global operations of foreign MNEs, rather than in the effectiveness of its own national industrial base (and, of course, especially that part of it taken by indigenous enterprises), it may cut back severely on such funding. If many leading industrial countries gain the same viewpoint and implement the same response, the result could be a significant global decline in basic research and a slowing of overall technological progress. Second, if a government notes that its own MNEs are performing large amounts of their R&D overseas and that the global efficiency of their operations gains from it, with ultimate benefits to the home country, the government may decide to take a free ride on this situation and cut back support for domestic science. Again, if a number of countries perceive the situation in this way, the overall effect on R&D inputs and scientific progress may be negative.

The three issues we have just outlined indicate that the globalization of the technological perspectives of MNEs possesses the potential to improve the ability of science to contribute to an enhanced generation of economic welfare, but they also suggest that uncertainties and suspicions about the way such efficiency gains are distributed may ultimately compromise the effectiveness of the process, through technological protectionism or even opportunistic behavior by governments. To analyze the nature of the implications of the globalization process for individual host countries, therefore, does not imply an ideological constriction of the conceptual scope, but rather that securing all its possibilities does require acknowledgment and understanding of different dimensions of interest among contributing parties.

The host-country implications of R&D by foreign firms may be seen at two distinct levels. The first of these relates to its effects on the short-term and medium-term performance of the industrial sector within which the foreign subsidiaries operate. The second concern is with the longer-term development of the host-country science base itself, following the interjection of foreign-controlled R&D activity.

Lines of analysis developed throughout this book clearly indicate the positive potentials at the first of these levels. The MNEs bring the technology underlying important new group-level product concepts and then use local creative inputs in a complementary manner to achieve the most commercially effective application of it to the markets to be supported by local production facilities. The new technology does not then merely extend the scope of industry but, through the involvement of local science, does so in the most effective and responsive manner possible. Local consumers can then benefit from new or enhanced-quality products, with the expanded demand then, it is hoped, generating expanded output, employment, exports, and government tax revenue.

However, some attention also needs to be paid to the opportunity cost of the employment of capable local scientists in foreign R&D units. This cost will be low if the scientific labor market is slack and these personnel would not have found work (or work reflecting their full abilities) in local firms. Indeed, if these researchers would have migrated in the absence of employment opportunities in a foreign-controlled laboratory, a type of external or spillover benefit may often be detected resulting from their retention in the wider science base. Thus particularly capable scientists often play valuable leadership roles outside of their direct local employment by serving on industry committees, education advisory panels, and other such bodies and generally contributing to the wider health of the scientific community. If employment in foreign MNEs' labs enables these researchers to continue to play this role, a significant benefit may be considered to accrue.

If the scientific labor market is tighter, then it is likely that recruitment of local scientists into foreign-controlled laboratories will have an opportunity cost in terms of slowing the technological development of indigenous enterprises. Even if the MNE subsidiary does make more effective and productive use of the talents of these personnel, this situation may still be contentious if there is a political or social predisposition to place a premium value on the growth of indigenous firms. Whether or not that attitude is considered valid is, however, beyond the concerns of this book.

Alongside the efficiency with which a country's current scientific capacity is used, we have also suggested that it is important to assess the way in which its competence and scope develop over time. One possible criticism of the operations of foreign MNE laboratories in this context is that where these units focus on IIL-type work, as part of a wider group-level program, the results will have no direct and immediate application to the technological capabilities of the country's industry (including that part controlled by the MNEs that own such laboratories). This situation may then resemble an "internal brain drain," with the work of parts of the science base having no more local relevance (either to technology or commerce) than if the personnel involved had actually migrated. It could also then be suggested that this form of IIL work may sometimes create the type of gap between a country's science and its industry that an efficient LIL seeks to fill.

However, the involvement of the work of foreign-controlled R&D laboratories also possesses the capacity to enhance the longer-term potential of a country's scientific community by providing it with wider operating perspectives and thereby broadening its effective scope. This enhancement can come both through the employment of local scientific personnel in the labs and through the collaboration of local research institutions (universities and so forth) in research programs implemented by the MNEs. At least two types of valuable learning may be acquired by these personnel or institutions in these associations. First, because they are likely to be applying their own abilities to the existing scientific background of the MNE, either to implement it in a particular market environment (LIL work) or to help extend it through additional complementary research (IIL work, perhaps involving collaboration with other local institutions), the host-country scientists will extend their knowledge of the technology relevant to their areas of expertise. Second, the local scientists may observe and learn new organizational perspectives by operating within the ambitious programs of MNEs. These perspectives may include improved viewpoints on the commercial application of technology or clearer perceptions of the conception, organization, implementation, and evaluation of the more precompetitive programs of research.

# References


Bartlett, C. A., and Ghoshal, S. 1989. *Managing Across Borders—The Transnational Solution.* London: Hutchinson Business Books.

———. 1990. Managing Innovation in the Transnational Corporation. In *Managing the Global Firm,* ed. C. A. Bartlett, Y. Doz, and G. Hedlund. London: Routledge, 215–55.

Bonin, B., and Perron, B. 1986. World Product Mandates and Firms Operating in Quebec. In *Managing the Multinational Subsidiary,* ed. H. Etemad and L. Séguin Dulude, 161–76. London: Croom Helm.

Cantwell, J. A. 1991. The International Agglomeration of R & D. In *Global Research Strategy and International Competitiveness,* ed. M. C. Casson, 104–32. Oxford: Blackwell.

———. 1993. *Multinational Corporations and Innovatory Activities: Towards a New, Evolutionary Approach.* Discussion Papers in International Investment and Business Studies, series B, vol. 5. University of Reading, Department of Economics.

Caves, R. E. 1971. International Corporations: The Industrial Economics of Foreign Investment. *Economica* 38:1–27.

Chesnais, F. 1988. Technical Cooperation Agreements between Firms. *STI Review,* no. 4:51–120.

Cohen, W. M., and Levinthal, D. A. 1989. Innovation and Learning: The Two Faces of R&D. *Economic Journal* 99:569–96.

Cordell, A. J. 1971. *The Multinational Firm, Foreign Direct Investment and Canadian Science Policy.* Science Council of Canada, Special Study No. 22. Ottawa: Information Canada.

———. 1973. Innovation, the Multinational Corporation: Some Policy Implications for National Science Policy. *Long Range Planning* 6, no. 3:22–29.

Crookell, H. 1986. Specialisation and International Competitiveness. In *Managing the Multinational Subsidiary,* ed. H. Etemad and L. Séguin Dulude, 102–11. London: Croom Helm.

Crookell, H., and Caliendo, J. 1980. International Competitiveness and the Structure of Secondary Industry in Canada. *Business Quarterly* 45, no. 3:58–64.

Dunning, J. H. 1993a. *The Globalisation of Business.* London: Routledge.

———. 1993b. *Multinational Enterprises and the Global Economy.* Wokingham: Addison-Wesley.

———. 1994. The Strategy of Japanese and US Manufacturing Investment in Europe. In *Does Ownership Matter? Japanese Multinationals in Europe,* ed. M. Mason and D. Encarnation, 59–86. Oxford: Clarendon Press.

Dunning, J. H., and Cantwell, J. A. 1991. Japanese Direct Investment in Europe. In *Multinationals and Europe 1992,* ed. B. Burgenmeier and J. L. Mucchielli, 155–84. London: Routledge.

Dunning, J. H., and Pearce, R. D. 1985. *The World's Largest Industrial Enterprises 1962–1983*. Aldershot: Gower.

Graham, E. M. 1992. Japanese Control of R & D Activities in the United States: Is This a Cause for Concern? In *Japan's Growing Technological Capability,* ed. T. S. Arrison, C. F. Bergsten, E. M. Graham, and M. C. Harris, 189–206. Washington: National Academy Press.

Granstrand, O., and Sjolander, S. 1992. Internationalisation and Diversification of Multi-technology Corporations. In *Technology Management and International Business,* ed. O. Granstrand, L. Hakanson, and S. Sjolander, 181–207. Chichester: Wiley.

Hagedoorn, J. 1993a. Understanding the Rationale of Strategic Technology Partnering: Interorganisational Modes of Cooperation and Sectoral Differences. *Strategic Management Journal* 14:371–85.

———. 1993b. Strategic Technological Alliances and Modes of Cooperation in High-technology Industries. In *The Embedded Firms: On the Socioeconomics of Industrial Networks,* ed. G. Grabher, 116–37. London: Routledge.

Hagedoorn, J., and Schakenraad, J. 1991. *The Role of Interfirm Cooperation Agreements in the Globalisation of Economy and Technology.* The MONITOR-FAST programme. Prospective Dossier No. 2: "Globalisation of Economy and Technology," vol. 8. Commission of the European Communities.

———. 1993. Strategic Technology Partnering and International Corporate Strategies. In *European Competitiveness,* ed. K. Hughes, 60–86. Cambridge: Cambridge University Press.

Hakanson, L. 1981. Organisation and Evolution of Foreign R & D in Swedish Multinationals. *Geografiska Annaler* 63B:47–56.

Haug, P., Hood, N., and Young, S. 1983. R & D Intensity in the Affiliates of US-owned Electronics Companies Manufacturing in Scotland. *Regional Studies* 17:383–92.

Hood, N., and Young, S. 1982. US Multinational R & D: Corporate Strategies and Policy Implications for the UK. *Multinational Business* 2:10–23.

JETRO. 1991. *The 7th Survey of European Operations of Japanese Companies in the Manufacturing Sector*. Tokyo: JETRO.

———. 1992a. *JETRO White Paper on Foreign Direct Investment 1992—Summary*. Tokyo: JETRO.

———. 1992b. *The 8th Survey of European Operations of Japanese Companies in the Manufacturing Sector*. Tokyo: JETRO.

———. 1993. *The 9th Survey of European Operations of Japanese Companies in the Manufacturing Sector*. Tokyo: JETRO.

Kogut, B., and Chang, S. J. 1991. Technological Capabilities and Japanese Foreign Direct Investment in the United States. *Review of Economics and Statistics* 73, no. 3:401–13.

Linder, S. B. 1961. *An Essay on Trade and Transformation.* New York: John Wiley.

Mansfield, E. 1988. The Speed and Costs of Industrial Innovation in Japan and the United States: External vs Internal Technology. *Management Science* 34, no. 10:1157–68.

Mowery, D. C., and Teece, D. J. 1992. The Changing Place of Japan in the Global Scientific and Technological Enterprise. In *Japan's Growing Technological Capability,* ed. T. S. Arrison, C. F. Bergsten, E. M. Graham, and M. C. Harris, 106–35. Washington: National Academy Press.

Ozawa, T. 1979. *Multinationalism Japanese Style: The Political Economy of Outward Dependency.* Princeton: Princeton University Press.

———. 1991a. Japanese Multinationals and 1992. In *Multinationals and Europe 1992,* ed. B. Burgenmeier and J. L. Mucchielli, 135–54. London: Routledge.

———. 1991b. Japan in a New Phase of Multinationalism and Industrial Upgrading: Functional Integration of Trade, Growth and FDI. *Journal of World Trade* 25, no. 1:43–60.

———. 1992. Cross-investments Between Japan and the EC: Income Similarity, Technological Congruity and Economies of Scope. In *Multinational Investment in Modern Europe,* ed. J. Cantwell, 13–45. Aldershot: Elgar

Papanastassiou, M., and Pearce, R. D. 1990. *Host Country Characteristics and the Sourcing Behaviour of UK Manufacturing Industry.* Discussion Papers in International Investment and Business Studies, series B, vol. 2. University of Reading, Department of Economics.

———. 1993. Global-innovation Strategies of MNEs and European Integration: The Role of Regional R & D Facilities. In *International Business and Europe After 1992,* vol. 1 (proceedings of the 19th Annual Conference of the European International Business Association, Lisbon, December 12–14, 1993), ed. V. C. Simoes, 327–52. Lisbon: C.E.D.E.

———. 1994a. The Internationalisation of Research and Development by Japanese Enterprises. *R & D Management* 24, no. 2:155–65.

———. 1994b. Host-country Determinants of the Market Strategies of US Companies' Overseas Subsidiaries. *Journal of the Economics of Business* 1, no. 2:199–217.

———. 1994c. The Creation and Application of Technology by MNEs' Subsidiaries in Europe, and Their Role in a Global-innovation Strategy. In *The Changing European Environment* (proceedings of the 21st Annual Conference of the UK Academy of International Business, UMIST, March 25–26, 1994), ed. M. Yamin, F. Burton, and A. R. Cross, 774–812. Manchester: UMIST.

———. 1994d. Technology Sourcing and the Strategic Roles of Manufacturing Subsidiaries in the UK: Local Competences and Global Competitiveness. University of Reading. Mimeographed.

Pearce, R. D. 1989. *The Internationalisation of Research and Development by Multinational Enterprises.* London: Macmillan.

———. 1992. World Product Mandates and MNE Specialisation. *Scandinavian International Business Review* 1, no. 2:38–58.

———. 1993. *The Growth and Evolution of Multinational Enterprise.* Aldershot: Edward Elgar.

Pearce, R. D., and Singh, S. 1992. *The Globalisation of Research and Development.* London: Macmillan.

Porter, M. 1986. Competition in Global Industries: A Conceptual Framework. In

*Competition in Global Industries,* ed. M. E. Porter, 15–60. Boston: Harvard Business School Press.

Ronstadt, R. C. 1977. *Research and Development Abroad by US Multinationals.* New York: Praeger.

———. 1978. International R & D: The Establishment and Evolution of R & D Abroad by Seven US Multinationals. *Journal of International Business Studies* 9, no. 1:7–24.

Suzuki, Y. 1994. *Multinationalisation of Japanese Manufacturing: International Locational Practices.* Discussion Papers in International Investment and Business Studies, series B, vol. 6. University of Reading, Department of Economics.

UNCTAD. 1991. *World Investment Report: The Triad in Foreign Direct Investment.* New York: United Nations.

———. 1993. *World Investment Report 1993.* New York: United Nations.

Vernon, R. 1966. International Investment and International Trade in the Product Cycle. *Quarterly Journal of Economics* 88:190–207.

———. 1979. The Product Cycle Hypothesis in a New International Environment. *Oxford Bulletin of Economics and Statistics* 41, no. 4:255–67.

Westney, D. E., and Sakakibara, K. 1985. The Role of Japan-based R & D in Global Technology Strategy. *Technology in Society* 7:315–30.

White, R. E., and Poynter, J. A. 1984. Strategies for Foreign-Owned Subsidiaries in Canada. *Business Quarterly* 49, no. 2:59–69.

Yoshitomi, M. 1992. Macroeconomic and Schumpetarian Features of Japanese Innovations in the 1980s. In *Japan's Growing Technological Capability,* ed. T. S. Arrison, C. F. Bergsten, E. M. Graham and M. C. Harris, 98–105. Washington: National Academic Press.

Young, S., Hood, N., and Hamill, J. 1987. *Foreign Multinationals and the British Economy.* London: Croom Helm.

# Index


Bartlett, C. A., 39
Bonin, B., 36

Caliendo, J., 33
Cantwell, J., 13, 14, 21, 31, 40, 97
Caves, R. E., 30
Chang, S. J., 31
Chesnais, F., 21
Cohen, W. M., 31
Consumer tastes, 2, 9, 16, 29, 30, 39, 95
Cordell, A. J., 33, 40
Crookell, H., 33

Dunning, J. H., 9, 10, 12–14, 18, 21, 27–30

Efficiency-seeking MNEs, 30, 32, 35, 36
European Community/Union, 3–7, 15, 16, 32, 77, 86
Exports, 8, 9, 13–15, 28, 39, 49, 85, 86, 95

Foreign direct investment, 3–7, 9, 10, 12, 27–30, 32
France, 11, 12, 54, 59, 61, 68

GATT, 32
Germany, 4, 12, 16, 19, 54, 59, 61–65, 68
Ghoshal, S., 39
Global innovation strategy, 1–3, 22, 37–42, 61, 69, 79, 81
Graham, E. M., 97
Granstrand, O., 37, 38, 49, 50

Hagedoorn, J., 21, 22, 24, 25, 56, 66, 67
Hakanson, L., 40
Hamill, J., 33
Haug, P., 40
Heckscher-Ohlin theory, 7
Hong Kong, 7
Hood, N. 33, 40

Independent laboratories, 81, 91
Industry funded laboratories, 81, 82, 92
Internationally interdependent laboratories, 42, 45–47, 50, 59, 74, 75, 91, 100
Italy, 68

JETRO, 3, 4, 6, 10–12, 15–17, 59–62, 69–75

Kogut, B., 31

Levinthal, D. A., 31
Licensing, 13, 14, 31
Linder, S. B., 29
Locally integrated laboratories, 41, 42, 45–47, 59, 60, 74, 75, 78, 91, 100
Local production conditions, 2, 29, 34, 35, 95

Mansfield, E., 31
Market expanding FDI, 29, 30
Market seeking MNEs, 28–30, 32–34, 39, 69, 72
Mowery, D., 67, 68
MITI, 9, 10, 67, 68

Netherlands, 4, 54

Oligopoly, 29
Ozawa, T., 7–9, 29

Papanastassiou, M., 1, 28, 38, 45, 85–92
Patents, 18–22, 52–55, 61–65
Pearce, R. D., 1, 18, 28, 32, 34, 36, 38, 40, 42, 44–46, 50–52, 86, 90, 97
Perron, B., 36
Pollution, 8–10
Porter, M., 34
Poynter, J. A., 33
Product adaptation, 29–31, 34, 35, 39, 41, 72, 78, 79, 89, 90, 95
Product development, 1, 2, 14, 30, 36, 39–41, 69, 71, 72, 74, 79, 83, 87–89, 91
Product mandate subsidiaries, 35, 36, 40–42, 69, 87–89, 91
Protectionism, 1, 9, 15, 29, 32, 34

Rationalized product subsidiary, 34, 35, 41, 87, 89
Research joint ventures, 68
Resource seeking MNEs, 27, 28, 33, 35, 36, 69, 72
Reverse engineering, 31
Ricardo-Hicksian trap, 8, 9
Ronstadt, R. C., 40–46

Sakakibara, K., 31
Scale economies, 8, 29, 30, 35
Schakenraad, J., 21, 22, 24, 25, 56, 66, 67
Singh, S., 42, 44–46, 50–52
Sjolander, S., 37, 38, 49, 50
Spain, 12, 59
Strategic asset seeking MNEs, 30, 31, 33, 36
Strategic technology alliances, 17, 21–23, 26, 31, 55–57, 65–67
Support laboratories, 40, 41, 44–47, 59, 60, 74, 75, 78
Suzuki, Y., 86

Taiwan, 7
Teece, D. J., 67, 68
Thailand, 7
Trade friction, 8
Truncated miniature replica subsidiaries, 34, 41

U.K., 3, 4, 11, 12, 19, 43, 46, 54, 59–65, 68, 75–77, 79, 80–82
UNCTAD, 3–7, 10
University laboratories, 81, 82, 91, 100
U.S., 3–7, 18, 19, 23, 26, 31, 43, 46, 54, 62, 63, 65, 68

Vernon, R., 39

Westney, D. E., 31
White, R. E., 33

Yoshitomi, M., 18, 19
Young, S., 33, 40